Life as a Geographer in India

This is the first book on the theme of the life of a geographer in India. The author introspects on her own experiences and engagements with the discipline and explores the life and works of over 24 other geographers from India.The volume documents and acknowledges the commitment of geographers to life, teaching and the subject of geography. Collectively these provide an insight into the growth and diversification of the discipline in the country. The book offers critical perspectives on the changing disciplinary practices within the field of geography by highlighting the major achievements and teaching methods of geographers. It brings attention to the diverse interests, themes and problems in geography pursued by these geographers while also influencing the lives of other researchers and professionals.

This book will be of immense interest to students, teachers and researchers of geography and social anthropology and readers interested in the lives of these influential educators and academicians.

Anu Kapur is Professor of Geography, Delhi School of Economics, University of Delhi, India.

Life as a Geographer in India

Anu Kapur

Routledge
Taylor & Francis Group

LONDON AND NEW YORK

First published 2021
by Routledge
2 Park Square, Milton Park, Abingdon, Oxon OX14 4RN

and by Routledge
52 Vanderbilt Avenue, New York, NY 10017

Routledge is an imprint of the Taylor & Francis Group, an informa business

© 2021 Anu Kapur

British Library Cataloguing-in-Publication Data
A catalogue record for this book is available from the British Library

Library of Congress Cataloging-in-Publication Data
A catalog record has been requested for this book

ISBN: 978-0-367-68688-8 (hbk)
ISBN: 978-0-367-71378-2 (pbk)
ISBN: 978-1-003-15055-8 (ebk)

Typeset in Sabon
by SPi Global, India

Contents

Introduction

I have often wondered about the ways in which geographers in India at large are pursuing their profession. How have they acquired the requisite learning and skills to grow the profession? Why have they come to acquire specializations in certain domains? What challenges have they faced? What does their narrative unfold about the times and location of their work? A fraternity hones a discipline. To know about one's members is an important dimension of concern for the subject. This theme somehow seems to have escaped the attention of geographers in India. It led me to ask where one could gain insight into the life of geographers in India.

Websites of the departments of geography in universities today carry brief professional biographies. The write-ups are signboards for universities to flag their faculty. They are designed to a specific template provided by the institutions and are akin to curriculum vitae where facts are listed without any reference to how and why these came to acquire their present form.

A bio sketch on book jackets is a second place that describes academics – including geographers – providing a pen picture of the author or editor of a book. In the case of the latter, a brief write-up on each contributor is given. This write-up is a one-paragraph description of the professional identity of the geographer. Although written by the author, publishers tend to reword it to suit their marketing agenda.

Obituaries are a third avenue for knowledge about the life of geographers. Almost all geographical journals in India have a section dedicated to the cause of providing a brief biography of deceased professionals. These are written in the nature of a eulogy for one's colleague, teacher or peer. Obituaries published in professional journals in India are few in number. Take, for example, the National Association of Geographers, India (NAGI) the largest professional body of the country; its journal, for the 10-year period 2009–2019, carried only 20 obituaries, most of which were less than a page and a half. The number is dismal not because editors are selective, but because few have bothered to take the time to write a tribute for a colleague with whom they have occupied offices, across the same corridor, for decades together. This and other such reasons means that even in obituaries the documentation of the lives of geographers in India remains limited in number.

Festschrifts are a fourth outlet which make available life sketches of select geographers of India. *Festschrift* is a German word meaning a collection of writings, which follows an old academic tradition where a group of authors write articles to create a volume in honour of a colleague or mentor. Since a *festschrift* is generally for those who have held important positions, it is not difficult for admirers to collaborate and contribute. Such an edited volume is usually published on the occasion of retirement or an important anniversary of the geographer in question. The introductory chapter of a *festschrift* carries an essay on the life and works of the scholar being honoured. There are a handful of *festschrifts* in India and these are gaining in popularity.

In this book I have documented the biographies of 24 geographers from India. Along with these biographies I decided to add my own autobiography, bringing the tally to twenty-five. Why have I written about the lives of geographers in India? While a brief on geographers can be read on websites, author blurbs, obituaries and *festschrifts*, there is a lack of any bio-bibliographies or archives on geographers in India. This book titled *Life as a Geographer in India* will fill this vacuum. By virtue of becoming the first book on the theme of geographers in India the book cashes in on first ever which, while a tag which has instant appeal, was nonetheless not enough reason to draw me to this theme.

The more important stimulation was that I find life-writing a precious vehicle of communication. It allowed me to relate and understand the forces at play that build a profession. For me, the writing of life was both inspirational and therapeutic. The translation and interpretation of circumstances and events, including my own, allowed me to connect with the joys and trials which punctuate our journey as professional geographers.

I feel strongly that the members of one's academic community are as important as the discipline one studies. A study of the life of geographers provides insights into how the discipline evolved and spread. It is professionals who shape the direction, destination and destiny of a discipline. In continuity with the motivation is also a realization that students of geography in India know little about geographers of their own country. Drawing from our colonial antecedents, the education of geography in India is heavily influenced by the West. Students of geography in India are exposed to lives of geographers in the West, especially, the U.S., the U.K., Germany, Sweden and France. This is also because these are countries which have taken the pains to document the works and worth of their geographers. Whereas it is certainly worthwhile to be acquainted with geographers across the globe there is no denying that it is equally important to understand one's very own in India. For students pursuing this discipline in India the life of geographers on home turf would definitely be more relatable and useful. The familiarity of departments, the opportunities available and explored by geographers in India could be valuable sources of information and direction for Indian students. Many of the geographers whose life is documented in this book have studied in leading universities in the West. Their write-ups could provide scope to understand how scholars have carried the ideas of their alma maters to establish the subject in their native country. The diaspora of India spreads far and wide into every country of the world, with many among the teachers, scholars and students engaged with geography originating from India. Their desire to know about geographers back home makes the writing of life write-ups of Indian geographers expansive in scope. In a globalizing world, a reading on life of geographers in India could help forge a bond between communities across borders and boundaries. This work could well enthuse some courageous soul to attempt a spatio-temporal comparison of the life of geographers across the world!

Among the geographers in India none until yet have documented the life of 25 geographers. As all write-ups in the present book are authored by me, this brings a uniformity in approach, content and context. The book looks at how these professionals have evolved, expanded and found their feet in the subject of geography.

Hailing from different parts of the country, the geographers in this book are from the most diverse backgrounds imaginable and are spread across major universities in India including the University of Delhi, Jawaharlal Nehru University, Aligarh Muslim University, Panjab University, University of Allahabad, University of Hyderabad and the universities of Calcutta, Bombay and Madras, among others. This pan-India selection means the book provides scope for gaining insight into how different habitats have shaped what these geographers have done and how they, in turn, have influenced the departments that have nurtured them. Many among this list of geographers are the founders of the discipline in India. This means that their life unfolds the stepping stones of how the discipline of geography evolved to its present status in India.

Many started out as schoolteachers, have served in different offices for the government of India ranging from the Planning Commission and Town and Country Planning Departments to the Census of India and most, if not all, have traversed many departments. Almost all have taken prestigious scholarships and fellowships. Several have studied abroad and some have served at foreign universities. Most of these have been highly influential teachers who largely held positions of presidents of various geographical associations in the country. Some have held vice chancellor positions. Their lives capture the varying impacts of these experiences.

The book includes both genders. While there are only two women compared to 23 men, the important feature is that women geographers are not absent in the scheme of things. The first generation of geographers in India form the bulk of the selection of 25, though a small number of the second generation pepper the discussion.

The geographers in this group include agricultural, urban, population, political and environmental geographers, geomorphologists and cartography specialists, among others. Many have been pioneers in their discipline's specialization and thus played a critical role in establishing schools of geography, adding to the colour and personality in geography departments across India.

These 25 geographers have been teachers, supervisors, mentors, collaborators, editors, authors, reviewers, examiners and many other roles to hundreds of geographers across India. I am sure that the published works of these 25 geographers have informed the teaching and research of scores across India. Pursuing diverse interests, themes and problems in geography, they all have added in one way or the other to the growth and expansion of the discipline in India.

I would be doing grave injustice to a large community of geographers if I made a claim that the geographers in my selection are the only or best representatives of the discipline in India. This would be a false assertion. I am well aware of numerous geographers whose life should be placed on record. This book would make a major contribution if it inspires others to take responsibility to document the lives of all those worthy geographers they feel deserve attention.

The first department of geography in India was established at Aligarh in 1924. Today we have over 150 university departments offering geography in India. In this almost 100 years, many geographers have served and strived for the discipline. The life of geographers carries the toil and triumph of their works and contributions. It matters little if they have been doyens, leaders, followers, laggards or drags. Embedded in the story of their professional life are anecdotes which enrich the discipline. I am certain many geographers carry interesting narratives and experiences; even their autobiography would contribute to the discipline's cause. Although I would have liked to, I simply am not equipped to write about all geographers in India. So how did I select this sample? This handful fell into my kitty as I had the opportunity to personally interact with them at some duration of time and this gave strength and confidence to write about their life.

These 25 lives were written at different periods of my life and for different reasons. According to the time and nature of interaction they differ in their substance and style of writing. These distinctions allow the write-ups to be classified into three categories: LifeContours, LifeScapes and LifeSpace. The choice of using geographical terms: contour, scape and space, is not an aberration but fits the approach that has been adopted to describe the write-ups. The design to write the life write-ups adopting different methods will be clear when I share how, when and why these geographers came into my life. It would be vain to conclude that there is only one lens to capture life; on the contrary, there are many possible ways to comprehend and represent a geographer's life. These have been mine.

LifeContours of 22 geographers in India

While depicting the life of 22 geographers I have followed a style similar to the drawing of a contour, hence the name, LifeContours.

We all know that a contour is a standard cartographic convention used to represent an elevation of land. The origin of the concept of contour lies with Charles Hutton, a British mathematician whose ambitious 1778 survey of a Scottish peak called Schiehallion marked its first known use. Hutton invented contour lines not in order to represent Schiehallion's 3,500 feet height, rather he created the contour to calculate the weight of the mountain. The contour was devised to solve a problem which had been taxing him for long and was part of a project to calculate the weight of the Earth using Newton's theory of gravitation. I am digressing to share this detail because in spite of teaching the drawing and interpretation of contours for 30 long years, it is only now that I have stumbled upon this information and shared it in case the reader is as naïve as me!

Arriving at the core issue, the question is why I call this group of vignettes LifeContours.

While writing *Voice of Concern: Geography in India* an important component was the inclusion of the presidential addresses of 22 geographers from the National Association of Geographers, India. I decided to prefix a brief biographical note about these presidents as a prelude to the presentation of these addresses.

A contour is an isoline joining places of equal elevation. A common feature of all these 22 geographers was that all had reached the rank of president of this association and this inspired me to call these write-ups LifeContours.

A second dimension which reflects this isoline is that all these were written within a short span of six months. This is because the book *Voice of Concern* had to be released in December of 2001 at the Annual Meeting of the National Association of Geographers, India at Sagar in Madhya Pradesh. I had begun to write this book in the middle of 2001 and therefore only six months were available to complete life write-ups.

That I had no professional contact with any of these geographers becomes the third shared feature of this group. None among these 22 geographers had ever been my teacher. I had never collaborated or attended a conference with any; nor did I ever publish a joint paper or write a review on any of their works. Apart from an occasional official meeting or an assignment for an exam I had hitherto not met any and thus cannot claim to know any of them. In this context I decided to get to know them through in-depth interviews.

A fourth common attribute of these 22 accounts is that the primary method of collecting information rested on interviews. Guided by a few close-ended and more open-ended questions, the insights of each life were collected over a minimum of three to four interviews per geographer. The interviews stretched over hours and were not easy to conduct. I had to keep changing the questions as I went along, and I stopped only when the sharing began to be repetitive or to confirm a line of thought. Moreover, even though I had decided that my write-up would be brief, in order to carve the critical details I chose to not focus on only one segmented portion of their life. Only by attempting to grasp a life holistically can the underlying process and pattern be discerned.

This blueprint meant that, with the exception of just two, I personally interacted with all these geographers. The untimely demise of Moonis Raza and V.L.S. Prakasa Rao meant that I missed meeting with them. So, instead, I met with Mehdi Raza, Moonis Raza's younger brother and V. Raghavaswamy, Prakasa Rao's son, to provide clarifications and details.

Apart from these two geographers due to their passing, all the remaining 20 geographers went out of their way to accommodate my request for an interview. It is difficult to forget that G.B. Singh drove all the way from Patiala to Chandigarh in response to the phone call. G.S. Gosal extended an invitation to his residence and allowed questions to spill into sessions that spanned across two days. In Delhi, despite a troublesome spondylitis, K.V. Sundaram sat through hours in all patience, R.P. Misra was his unhurried self over a long discussion at his residence; Aijaz Ahmad in Gurgaon postponed his appointment to the doctor when he realized I was not in the mood to wrap up; S. Chakraborty extended his stay in Delhi and devoted two days to share his story; L.S. Bhat drove all the way from his home to the Department of Geography, DSchool, with a promise of a meeting over lunch that ended up lasting until dinner. Pathak's response to my telegram led him to cancel appointments scheduled in Delhi; he was full of answers right up until he boarded the train back to Calcutta. A drive to Aligarh saw Mohammad Shafi; he had taken care to reserve an undisturbed block of time in his chambers at the department. Patna is a city which houses two presidents and here I interviewed P. Dayal and R. Ram at their bungalows. Ram's desperate search for his presidential address did not meet success and remains a lacuna in this book. Surrounded by a hospitable family and four dogs, P. Dayal's formidable age did not thwart his alert response to many an incident from the past.

Mahadev resides in Chennai, Kayastha in Varanasi, Alam in Hyderabad, Vaidyanathan in Cuddalore, Das in Guwahati and Arunachalam in Mumbai. Geographical distances were overcome by enrolling this group of presidents in conversation over the telephone. Health impairments in the case of Mahadev, Das and Kayastha meant that questions had to be asked over short sessions and at times members of their family were interpreters – but the willingness to answer was universal. K.N. Singh was the only geographer who was not in India, he was in Ethiopia in a professional capacity. His interviews were conducted over email and long-distance phone calls! I admire the universal enthusiasm of all these geographers while sharing their life trajectory with me.

By virtue of holding the post of president, each came from the position of knowing a lot about geography and geographers in India. I guarded their sharing and refrained from cross-checking with colleagues or their students. The challenge was not to compare one with the other and I wanted to form my own perception about each. It is for this reason that the LifeContours are drafted in my words and not as a direct narration by the interviewees.

In hindsight, I took what they said at face value. I mentioned the number of books and articles each wrote, but chose not to assess their published works. This is because I could simply not have done justice to a critical review of their works because time was short and these 22 geographers had many publications. I was cautious and chose to be silent where I did not know enough.

I decided to keep these 22 write-ups brief and thus most of them fall in the range of 1,000 to 1,500 words. Like a contour, I opted to keep the write-ups simple and crisp, where the configuration tells one if it is a hill or valley, the closeness of contours indicates a steep slope and the irregular contours represent a rugged terrain.

In complete contrast with the 22 LifeContours is the genre of LifeScapes to which four geographers belong.

LifeScapes of four geographers in India

LifeScape is a narrative that details the intersection of events in an individual's emotions, socio-economic circumstances, and behaviour over their lifespan. There is little doubt that as a geographer, I borrowed the lead to use this word from the concept of a landscape and tried to capture the life of the geographers with reference to both the scene and scenery.

The nature of this approach restricts the number of lives I could write about. In contrast to the 22 LifeContours only four fill this basket. Unlike the LifeContours all of which were written within six months, the LifeScapes individually took over six months and moreover I wrote these in different years of my life. Here is the sequence: S.G. Burman in 1995, C.P. Singh in 2000, G. Krishan around 2005 and K.V. Sundaram in 2010.

I did not deliberately pick to write about these lives at an interval of one every five years. I could not have in any way. This will become clear when I elaborate how these four geographers came to be included in my selection.

Both S.G. Burman and C.P. Singh were my teachers in the Department of Geography, DSchool, the former my doctorate supervisor too. S.G. Burman died in 1995 and C.P. Singh in 2000. Though I was sad at their death, I was filled with gratitude for the role they had played in my life and so a few months after their demise I wrote their life stories.

The Department of Geography had formed the Association of Geographers (AGS), Delhi in 1979. The association publishes monographs under the title *Vasudha* (the earth). Since S.G. Burman was a founding member and held the post of editor and president of the association, I thought it would be befitting to publish her life write-up. When the turn to publish the write-up on C.P. Singh arose, I again knocked on the doors of AGS. C.P. Singh too had been a founding member and the year he died, he was the serving president and chairperson and the cardiac arrest that led to his instant death happened in his office in the Department of Geography. All these are good enough reasons for AGS to consider a life write-up on its own members. To my utter dismay, the editorial board of the AGS outright rejected the publication. It was not even sent for consideration for a peer review.

When I submitted my manuscript to AGS the latter did not have any other script that could pose as competition. The blatant rejection of C.P. Singh's LifeScape confirms what we all know: some of our colleagues can turn scathingly uncouth at critical moments. I thought nothing could be so crass as to deny one's colleague a place he deserved. For days I must have carried a hurt look until B. Thakur a professor in the department, in kindness, took the manuscript from my hands and assured, ".. just leave the destination of this to me now."

In a few weeks I received a letter ... *Many thanks, I received your write up of Chandra Pal Singh. You have written a most interesting essay and I expect to have it published in Volume 23 of Geographers Bibliographical Studies. This is still some months distant. Did you know that O.H.K. Spate passed away some months ago?* It was signed Emeritus Professor Geoffrey J. Martin, 24 November, 2001.

I was familiar with Professor Martin's book, *American Geography and Geographers.* I also knew that *Geographers: Bibliographical Studies* which had accepted the publication of the write-up on C.P. Singh owes its origin to the Commission on the History of Geographical Thought created by the International Geographical Union (IGU). What I did not know is that the Commission on the History of Geographical Thought had been created in the IGU meeting at Delhi in 1968.

From their first volume in 1977 to its most recent in 2019, *Geographers: Bio-bibliographical Studies* has published the life of 700 geographers. It contains the write-ups of only three geographers from India: N.K. Bose, S.P. Chatterjee and C.P. Singh. We all are familiar with the landmark works of N.K. Bose and his contribution in the field of human geography while serving at the University of Calcutta. As for S.P. Chatterjee, he was founder of the Department of Geography, Calcutta University and founder of the National Atlas Organization – later renamed as the National Atlas and Thematic Mapping Organization and a recipient of the Padma Bhusan by the government of India in 1985 for his contribution to geography. There could not have been a more worthy site to put to rest the life write-up on C.P. Singh. The editors of bio-bibliography modified the text to adapt to their style of publication. Its original version is presented here. Let me not forget to mention that it was the thoughtful deed of B. Thakur which made its destination possible. It also is a truism that our colleagues can be most helpful and life continues with a learning to give and forgive.

An entirely different chain of events led to the writing on G. Krishan. I had not been his student and he and I had never researched together. I met him for the first time on 7 March 1996 at a seminar on Geography and Public Policy, organized by the Department of Geography, at Panjab University, Chandigarh. When G. Krishan retired from Panjab University in mid-2002, I received a call from his doctoral student and colleague, Surya Kant with a request to write the life and works for a *festschrift* on G. Krishan. I was keen to oblige.

G. Krishan was one among the 22 presidents of the National Association of Geographers, India and I had written a brief on him, yet I knew there was more to know and share, hence there is a write-up in both the sections of LifeScapes and LifeContours. A comparative reading will help readers grasp how the two categories capture the life of an academic.

An entirely different set of reasons led me to write on K.V. Sundaram. The latter had taken geography out of the ambit of the classroom to the corridors of policy and regional planning. Similar to G. Krishan, K.V. Sundaram had not been my teacher nor had I collaborated with him in any professional capacity. The events of why I set out to reach him are as follows:

When I was faculty at Kirori Mal, a constituent college of the University of Delhi, I was nominated as the student staff adviser. The post carries the responsibility of organizing *Geotime*, the annual festival of the Department of Geography in this college. For this occasion, I struck upon the idea of organizing a four-day seminar on the theme of Careers in Geography. To this end, I was looking for geographers who were employed in different professional capacities.

I succeeded in netting a cartographer, weather reporter, environmental journalist, tourism consultant, director of a mapping company, and even an officer from the Census of India, an organization in which geographers play a vital role in creating the Census of India's atlas. To complete the quorum, I desperately needed an urban and regional planner, without which I believed my seminar would be incomplete. This stemmed from the conviction that both urban geography and regional planning are an important component of the syllabus of a graduate studies programme and many students aspire to cut a career in this domain.

It is with this awareness that on a Monday evening sometime in November 1994, I knocked at the home of K.V. Sundaram, with a request to address my students on a career in regional planning. There was no looking back. K.V. Sundaram and I struck an immediate academic discourse; one which continued to blossom until the last days of his life.

By 2000, K.V. Sundaram had spearheaded the inception of Bhoovigyan Vikas Foundation, a non-profit organization, to promote the cause of Earth Sciences. In 2001, the Foundation bestowed a Leadership Award to four middle level geographers in India below 45 years of age. My name was among them, the other three being, A.C. Mohapatra, V. Raghavaswamy and S. Chattopadhyay. I was the youngest and felt delighted. The award carried a citation along with a cash prize. Perhaps out of gratitude and more out of my admiration for the contribution of K.V Sundaram that I wrote on life in 2010. It remains unpublished to date.

Even though I had personally known and interacted with all four geographers for over a decade, I found it difficult to join the dots of their life. It dawned on me that in our preoccupations, we often lose out on observing the important facets of the life of our teachers, professionals and dear ones.

LifeScapes were complex to draw. To gain insight, similar to drawing a landscape, I traced profiles of the origins, education, work and achievements of these geographers. The families of all four generously opened treasure chests filled with letters, invitations, awards and assignments. A careful read of their timetables, official notices and student reports in their official or other capacity told me the planning and effort that went into assignments undertaken. From their photographs I learnt where they went for holidays and what were the sites of fieldwork with their students. A browse through their personal library gave me a glimpse of their interests in wider reading.

The details of what shaped the life of these professionals were bolstered by discussions with students, colleagues, friends and family members. Based upon a rich archival collection and close interaction, the emphasis of these four write-ups is on how these geographers built a body of specialization within the framework of their situations and times. An effort is made to provide insight into their trials and tribulations and how they influenced and interacted with geography and geographers at large.

The narratives are dense and wherever necessary tables and graphical presentation are inserted to capture the ebb and flow of their life works. All four geographers provide startlingly different LifeScapes defined and guided by their own objectives and choices. But one fact threads them all and that is their passion for geography.

Writing about the lives of geographers unexpectedly reaped a bounteous harvest. I enjoy reading about people's lives of all kinds – in the writings of each is a textured diversity, a granularity of a kind that reveals much about life.

The research allowed me to host some leading geographers over a simple home-cooked meal, wherefrom emerged leisurely talks on geography and geographers that would never have been told in a public space.

The frantic collection of facts and figures, hours of interviews and flipping across letters and files turned into a source of contemplation. In my quietude, the writing of these narratives led me to ask what is the meaning of life as a geographer in India?

The variety in their personalities, thoughts and actions gave me a glimpse of how nuanced is the way of geographers in India. Geographers discussed not only their own lives, but also the habitat of their social, economic and political spaces. Their sharing helped capture the processes of change. I saw how each had contributed to help create the structure of the discipline we inherit today.

I learnt that geographers adopt different approaches in trying to achieve their professional goals. There are some who adopt the strategy of specialization; these choose to specialize on a single theme and thereafter produce a sizeable number of publications and supervise doctorates on a single theme, allowing them to expand their area of influence. By concentrating on and becoming an expert in a particular branch of the subject they are able to build schools of geography in India such as population geography, agricultural geography and urban geography, among others. A second approach is where geographers are quick to spread their presence by attending or organizing seminars and conferences and thus making their presence felt.

They often sit on as many committees as possible. This group relies heavily on teamwork, collaborative research and publishes with many authors and this approach also expands their area of influence. Through their channels of networks, they grow and gain a position within the discipline. A third group consists of geographers who look at what is new and popular in the discipline and seek to become trendsetters. To attain this mileage, they lecture on current themes and are continually sharing and connecting with professionals in the discipline. The fourth group comprises of geographers who choose an innovative approach and are keen to think through a new topic and become pioneers in the field. They tend to publish alone and to achieve their aspirations, they shy away from groups, the busy schedules of conferences, and travel and deadlines imposed by external agencies. Although the persona of an individual influences the mode and morality adopted, yet there is no clear way or path to realize one's professional goals and one may as well adopt different approaches as they journey on their way.

Since most of the geographers included in this book are in the 65 plus age group category with the majority above 70 years of age, I was able to gain insight into how they engage with geography in their later years – post retirement. I found that having risen to the highest career rung of becoming a professor, director, or vice chancellor, some slow down but with a deep sense of satisfaction of having completed most of their professional obligations. There are a handful of geographers who strive to collate their teaching notes or earlier publications and settle to compile their work. This group of geographers do not necessarily advance knowledge but organize and consolidate their own works. Freed from institutional responsibility, some geographers also shared their desire to write their seminal work. In my interactions with them, I found that family circumstances, health issues and untimely death can set limits to achieving their dream of finalizing and publishing their lifetime treatise. Such then is life.

At the time of life-writing about presidents of the National Association of Geographers, India, I had the opportunity to meet all the geographers but two who had died prior to 2000. Today I would have only had the chance to meet five of them. The other 17 have not lived to see 2021. While this could be seen as an important justification for why I have not updated the write-ups, the more important reason is that while years may add more publications, positions and honours, rarely do the persona and personality, the mores and methods of work of a professional show a marked change. The inherent attributes and specializations captured while drawing the Life-Contours or the LifeScapes continue to remain largely the same.

I will take the example of G. Krishan. I wrote his LifeScape in 2005. In the last 15 years his path has shown little shift or change. Today as Professor Emeritus, he is at the same Department of Geography, he continues to hold posts of presidents of various associations, he continues to win more awards and publish his favourite theme of development as a form of vitality for India. The number of conferences he attends has reduced but this is more to do with limitations of age rather than a matter of choice. His mannerism and gait remain the same: humble, helpful and kind. Are we all cast in a mould? Like a painting which one cannot be redone, so too LifeScapes and LifeContours cannot be rewritten, edited or updated. The design of the write-ups ensures their endurance.

That life is short and exits are quick, allows my modesty and humility to remain in check. I so wish now that I had written about many more geographers. As of now, I can truly say my life as a geographer would never have been what it is had I not mapped the life of geographers in India.

My LifeSpace as a geographer

My 60th birthday was on the 12 June 2019 and I stepped into the category of what people refer to as the young senior. Bracketed into this cohort, brought the realization that the formal door of professional engagements will soon draw to an end when I turn 65 in 2024.

Even if age is only a number, the reality that organizations have built-in expiry dates cannot be wished away. When I leave the portals of the university I have this strange feeling that I will not only miss the ambience, but I will miss the person I was at this time and place because I will never be this way again. For reasons unknown, instead of looking forward, the clock in me wound backwards and I turned my thoughts to the years that had gone by as a geographer.

I decided to do a sort of self-audit and put together all those hanging and nagging thoughts of what has been my life as a geographer in India. Moreover, since I had already written the life of 24 geographers in India, I felt confident – and proven publications helped me in this – that I could write my own life as a geographer. I also felt that writing a book titled *Life as a Geographer* would be rather incomplete if I did not write my own experiences. Furthermore, I feel I have had an interesting innings as a geographer and there are aspects worthy to share. I am aware that it is rare for the categories of autobiography and biography to be together in a single volume, but this is a different book; here all the biographies have been written by me and so they form a part and parcel of my life as a geographer. This means there is neither conflict nor contradiction. Besides, each life is different, that of 25 geographers opens the scope for diversified learning about geography and the geographer. Writing my life as a geographer was not as easy as I had anticipated.

At first it seemed daunting and there were days when I would spend hours staring at the screen of the laptop and would close the Word document without having written a single word. Looking back at 60 years, a lifetime gone and putting all this down in 100 pages was not as easy as I had initially thought. I know that autobiographies suffer from issues of memory, subjectivity and attachment to certain dimensions as well as rigid perspectives. Even though I tried to avoid them, I am sure mine too carries shades of these.

I purposely bypassed negative spaces which take their toll on life and chose not to venture into dark and dingy alleys. I balked several times and often felt at a roundabout, where I needed to make a choice of which direction I should traverse and the event I should write about.

The reason for opting for the word space in the caption of this autobiography is deliberate. Whereas in LifeContours I was drawing a silhouette of the geographer and in LifeScapes I was going the extra mile to fill in the details, in my own write-up, I stepped forward as an outsider looking at the insider and so needed the word space in my life.

This space provided multiple interactions to emerge. Who am I in life? What has life as a geographer meant to me? How have I worked my academics? Why did I make certain professional decisions? How did I hone my ideas and manifest my thoughts? The autobiography carries a section titled, "then and now" where I share my perceptions of the changing face of geography and academia in India.

Rooted within the geography *of* India, the life write-ups tell us about geography *in* India. It could well be the other way round, rooted within geography *in* India the life write-ups tell us about the geography *of* India. Lives of others are mirrors of our own life.

I hope you will enjoy reading *Life as a Geographer in India* as much as I have writing about these lives for you.

Part I
LifeSpace

1 My LifeSpace as a geographer

While my great, great paternal grandfather was hewing wood on the banks of the River Jhelum in undivided India – now Pakistan – and aspiring to set up his own business, the British rulers of India were debating on an education policy which would serve their purposes. In 1835, British historian and politician Thomas B. Macaulay presented his *Minute on Indian Education* that sought to establish the need to impart English education to Indian natives. Thereafter, the British laid the foundations for the first three modern universities in India in the presidencies of Bombay, Madras and Calcutta in 1857. But for nearly 50 years, these universities were merely examining bodies with no teaching or research. It was only with the introduction of the Indian Universities Act of 1904 that the possibility of postgraduate teaching and research in both the humanities and the sciences arrived in the country. Meanwhile, colleges multiplied from 25 in 1885 to 190 in 1901. Geography, however, had no place on any campus.

Even though geography was taught at the University of Oxford in 1887 and the University of Cambridge the following year it was not introduced within any academic setting in India. In contrast, the British were busy establishing institutions that were engaged in gathering geographical knowledge about India. The Survey of India was established to map and measure the country; the India Meteorological Department to understand its climate; the foundations laid for the Geological Survey of India to gauge its mineral resources; the Forest Survey of India set up to comprehend botanical resources; and the Census of India created to provide details about the country's population. To systematically document the knowledge of the territory of India, the British authored nearly 1,500 geographical dictionaries or gazetteers. All these feats were accomplished to gain knowledge of the land and people for the sole purpose of controlling and exploiting the resources of this country.

Geography was important, but up until when the first department of geography was inaugurated in Lahore at Panjab University in 1920 followed by a department at the University of Aligarh in 1924, there were no geographers in the country. By the time India achieved independence in 1947, she had 18 universities. Among these seven offered a course in geography: Aligarh (1924), Patna (1927), Madras (1926), Agra (1935),

Allahabad (1937), Calcutta (1946), Varanasi (1946) and Punjab (1947). The University of Delhi was established in 1922 but did not inaugurate a department of geography until 1959. While the country with a population of 330 million and an area of 3.29 million km^2 carried a lot of geography, lone geographers scattered across half a dozen universities, a few colleges and some schools was the state of teaching of this subject.

The year of India's independence, 1947, was also the year of the partition of the subcontinent into two independent nation states: India and Pakistan. The episode uprooted 15 million people, killing over one million and sparking the largest refugee movement in human history on religious lines. The Partition also became a decisive moment for my family.

My grandfather was forced to leave a flourishing business and a large estate of dozens of properties in the city of Lahore and Jhelum and resettle in Delhi. He was a visionary and had the political and economic resources to relocate the family to Delhi six months prior to Partition. This meant that while many families had to struggle with loss of life and separation, mine had crossed over safely with their belongings well before a new international boundary could tear apart the subcontinent of India into two separate countries. The distress and despair faced by people during Partition featured often in discussions in our home. Though we were materially well off, the elders of my family had a deep sense of humility and helpfulness towards their friends and relatives who had been displaced. I recall the gratitude many people expressed in later years, of how warmly they had been welcomed to stay in our home, while they searched for a livelihood or secured their own shelter.

To accommodate the students – including those pursuing geography – displaced from the Panjab University in Pakistan, the government of India started a college in Delhi in January 1948. The latter was called Camp College, a name which reflected the thousands of camps which were shelters for refugees across the city. Released in 1954, O.H.K. Spate's book, *India and Pakistan: A General and Regional Geography*, was a regional geography text capturing the emergence of separate nations which, to date, remains an unchallenged treatise.

To settle the upheaval created by the deluge of refugees, India's first prime minister, Jawaharlal Nehru set the goals of nation building, fostering the articulation of unity in diversity. To commence a programme of development, the government established the Planning Commission in 1951. With this initiation of the planning era the role of geographers was increasingly recognized by the national government to carry out regional surveys, and prepare resource inventory and mapping. The Planning Commission chartered a special focus on challenges in different regions such as the Himalayas, the desert, metropolitan cities and borderlands. Special programmes were rolled out for the development of hill areas, tribal areas and drought prone areas, among others. The emphasis on area studies gave geographers an opportunity to provide solutions to the problems that beset different regions of the country. Responding to the need for maps and atlases, the National Atlas and Thematic Mapping Organization (NATMO) was established in 1956 in Calcutta.

The Indian Council of Social Science Research was established in 1969 to incorporate social research into policy formulation. This gave geographers scope to seek funds to pursue research projects on various spatial issues and concerns. Earlier in 1961, the National Council of Educational Research and Training, New Delhi, had been created to support the production of textbooks on school education, including those for teaching geography.

I was born on 12 June 1959, and the birth of the Department of Geography at the University of Delhi was the following month in July 1959. Who would have envisaged that I would be spending over 30 years of my professional life at this department, which is a part of the prestigious Delhi School of Economics, University of Delhi? But before I was to heed the calling of this subject, I had an 18-year journey to traverse.

First 18 years

My family enrolled me for study at the Convent of Jesus and Mary, Delhi. Designed on the lines of the British schooling system, the all-girls school adhered to the Senior Cambridge Board regulations. My first learning of geography began at this school. The geography teacher, a petite nun, delivered a thoroughly interesting introduction to world knowledge in her soft Irish accent. The concept of regions was included and with it the lessons of differences between various parts of the world.

School geography in those days was about map locations and alongside we learnt how to draw and identify contours of hills and U and V-shaped valleys. The schoolbooks were a delight; maps and diagrams coloured in pastels and small questions and answers to help as guides. I can never forget the way the nun taught us how to draw the outline of the map of India; she went through every curve and curl of the coastline announcing at the most appropriate time when to take the pencil up, sideways or down – and as she spoke we got the map in place right – this proved most helpful as I could then draw the map freehand with ease when I began to teach the course on India at the university.

In school the emphasis was on the three Rs – reading, [w]'riting and [a]'rithmetic. I will never forget the stern wooden-faced teacher who trained us in cursive English handwriting. I laboured to acquire a meticulous script with a dot over the "i" and a cross over the "t," but with the arrival of word processors, the typing ruined it all. I also recall the elocution classes where every "v" had to be bitten and the "w" had to be round and the Wren and Martin was the bible for grammar. Within such rigour and emphasis, English became my first functional language and I lost my grip on Hindi, which became a struggle when I later had to teach a bilingual class at the university.

Bolstering my interest was a gift on my 12th birthday, from my uncle in London, of the yellow-bordered *National Geographic* magazine for five continuous years. These were a treat. I marvelled spellbound at the Earth scenery, for the photographs in the magazine were superb. I will never forget those beautiful maps tucked neatly in the flap on its back cover. Their intricate details and density of information bewildered my imagination. In one edition, instead of a map there arrived a blue plastic record. Those were the days when one listened to music on a record player. I was really excited, when I played the record, to hear the voices of blue whales singing in the North Atlantic Ocean. I was fascinated not so much with the idea of the music of the whales, but more with what geography as a subject could embrace.

Along with this magazine a childhood comic I enjoyed most was the string of *The Adventures of Tintin* in the Belgian Congo, Egypt, India, China, Tibet and the UK. Saving my pocket money, I bought dozens of Tintin books; what I admired were stories not about the real lands but the descriptions of territories of fantasy, ones that did not exist on this earth! The art of geography could be so fascinatingly used in stories for children through comics. By the time I was out of school the atlases, *National Geographic* and Tintin comics had filled my bookshelf.

School was a place of red blazer, white skirt and blouse, a lot of laughter and deep friendships and I am now able to reconnect to my childhood friends thanks to their efforts to search for me through social media platforms. On our meetups we, now in our sixties, feel like stars which had been scattered across the sky. Looking back at school, although I attained a first division and a high score in geography I could never ever imagine becoming an academic; my inclination was to sing and dance and enjoy the world rather than study it.

My family was not one that engaged in intellectual discussions and debates. Education was not a priority, especially among girls. Only one among my six *bhuajis* (aunts) completed her graduation, though each lived life well, are proof that wisdom is far more important than formal degrees. My father went to a school that was established by his father and so carried the name Hithkari School in sync with their business nomenclature Hithkari Bros Pvt. Ltd. I am told that the school still exists in the same location with the same name in Jhelum, Pakistan. My father tells me that the school bell would ring only after he and his brother and sisters reached school. 'What a privilege!' I thought. When I asked him about the geography he learnt in school he told me that it was taught along with history for the matriculation examination. He never seemed to have particularly liked this subject.

Though my father is a chemical engineer from the prestigious Delhi College of Engineering, it was my mother who drummed into us the merit of education as the route to a good and meaningful life. It is she who instilled my love for books. She also paints and the inculcation of an artistic bent of mind was most helpful when I set to design maps. Her three brothers had taken their degrees in the UK. She herself is a science graduate from the famous Isabella Thoburn College Society of the Methodist Church at Lucknow. All in all, British culture had a space in our traditional Punjabi family. Classics such as Shakespeare, Dickens, Hardy and Tolstoy along with the romantic trash of Barbara Cartland and Mills & Boon were read to my heart's delight. In addition, one also could differentiate the works of painters like Degas, Van Gogh, Picasso from the likes of M.F. Husain and Amrita Sher-Gil. As a family, we could be seen more regularly at the Kamani Auditorium, Shri Ram Centre for Performing Arts and the Siri Fort Auditorium attending concerts, plays and art exhibitions than at shopping plazas and cinema halls. The programmes were meticulously planned by my mother and we would wait eagerly for those special weekends. The enjoyment of good Punjabi food, music and dance was, of course, on the forefront.

All through my school, a dedicated *guru* (teacher) came home to train me in *kathak*, a classical dance form introduced in India during the Mughal period. That I had a talent for this performing art is evident from the Best Dancer medal bestowed on me when my school celebrated its golden jubilee. When my *guru* wanted to enrol me for a world dance troupe, my conservative family ensured that I would never see him again! Today the same family members are glued to the Dance India Dance shows and swoon at how talented India is, while back in the mid-1970s I was sternly advised that to make a life it is better to read and write than be a dancer. A brief I seemed to take too seriously as life would later show.

But no regrets, the days were joyful and good. I recall that for many continuous summers we were taken to Mussoorie, a hill station, about 300 kms to the north of Delhi. Escorted with the retinue of our housekeeping staff and maids, we would spend our summer in a palatial, rented, hill house, learning how to horse ride, skate and enjoy the nature's walk on the Camel's Back Road – a road which was given this misfit desert name because of a rocky outcrop in the shape of a camel's hump on one of its hairpin bends.

Material comforts were plenty, yet a sense of patriotism and sensitivity to those who were not so well endowed was always instilled in parental mentoring. Strange as it may seem, a love for India was enforced when we were taken to watch select Bollywood films which revolved around social themes. Films like *Shaheed* (Sacrifice) or *Upkar* (Goodness) demonstrated the role of patriotic leaders during the fight for independence, and others such as *Do Bigha Zamin* (Two Square Yards of Land), and *Roti, Kapada aur Makaan* (Food, Clothing and Shelter) which depicted the struggle over the bare necessities of life, were discussed at length to get a feel for the real India; its poverty and distress. I realize media plays a critical role in the shaping of young minds.

When war clouds gathered over India in 1962, 1965 and 1971, I was three, six and 12 years, respectively. I vividly recall the Indo-Pak War of 1971, as a time when windowpanes were covered with brown paper and we were instructed to stand in the corners of the walls when air raid sirens were heard. The family generously packed cartons with rations, clothes and medicines for the *jawans* (soldiers), something which my mother continues to do when any major disaster strikes the country. Through such examples we were moulded to modest living and I developed a bond with my country.

When my Senior Cambridge exams finished, I was sent to London as the much promised holiday for devoting my time towards study instead of pursuing dance. As luck would have it from there I was given the chance to travel to Nairobi, Kenya. You see, the latter was my mother's birthplace. Her parents had migrated to Africa in the 1920s when the British were laying roads and rail tracks in Kenya. My grandfather took a ship across the Indian Ocean in search of a lucrative livelihood.

Travel reveals geography

As for me it was a wonderful trip of contrasts: Delhi, London and Nairobi; from the land of the tropical monsoon, to a temperate country only to be transported to the savanna landscape of the Kenyan highlands. I was sensitized to not just the cultures of the different racial groups but also the stark disparity between the developed, developing and underdeveloped realms of the world. Kenya turned out to be everything I thought it would be and more. A visit to the Masai Mara tribe in southern Kenya, the delight of a road sign near Nanyuki marking the equator, the breathtaking beauty of Mt Kilimanjaro, the highest single free-standing mountain in the world, left as much an imprint as the River Thames, Oxford, Cambridge, the Tate Gallery and a visit to Shakespeare's family homes, including Anne Hathaway's Cottage. Considering my colonial background of education, all this seemed familiar and was fun to experience. The captivating contrasts of culture and landscapes watered the seeds of my passion for geography which had been sown during school. It was on this rather heady holiday that my mother's eldest brother reinforced in me the value of studies stating clearly that countries in the West had progressed because they gave priority to education. It was the acquisition and creation of knowledge that had bestowed on them the power to conquer the world and fight the miseries of poverty and material deprivation.

On returning from my first foreign trip, my traditional family packed me off to another convent refuge to pursue an undergraduate programme, this time Sacred Heart, Dalhousie. The aesthetics of its castle architecture and the beauty of the Himalayas were undoubtedly alluring and appealing but I soon realized that bending over the cross stitch, learning to waltz on the Blue Danube and icing a cake with flowerets was not to be my lifelong vocation! The academic vigour at Sacred Heart Convent, was feeble and I was quick to relocate. This time around I firmly announced I was going to study geography.

Geography for me

By the 1970s geography was embedded in around 50 departments across various universities of India, of which I chose the University of Delhi, simply because it was located near my home. My personal geography influenced the choice of where I would undertake my study of geography it seems!

In 1976, this subject was offered for study in only five colleges of the University of Delhi. These were Miranda House, Kirori Mal, Dyal Singh and the morning and evening colleges on the two campuses at Shaheed Bhagat Singh College. Among these, my family decided on Miranda House partly because of its prestige but more because it was an all-girls' college.

I have no definite answer why I chose geography except perhaps it was about the earth, its land and oceans, its people and cultures, its diversity and disparity and that I simply love maps. Geography to me came with a bouquet of learning about the landscape and environmental features, climate and climate change, sustainability and conservation, population pressure and movement, resource crisis and conservation, cities and villages, transport networks and trade patterns, geopolitics and geostrategy, vulnerability and disasters; name it and there was geography. As I studied, taught and wrote about geography, the subject became ever more interesting and fascinating. It kind of unfolded itself at every step in known and unknown ways and so held on to my interest and imagination. Exactly when geography became my vocation or a calling I do not know, but my years of formal learning went by more quickly than I would have envisaged.

As a student at Miranda House

Looking at the first class marks obtained by you as indicated in your application; we are sure that you would be able to get admission in our institution. Thank you for your keen interest to join Miranda House is the letter I received from Miranda House, 24 June 1976. This warm reception was overwhelming and would be unheard of today.

Miranda is a prestigious college. The foundation stone of Miranda House was laid on the 7th March, 1948 by Lady Edwina Mountbatten, wife of the viceroy of India. The then vice chancellor of the University of Delhi, Sir Maurice Gwyer, named the college Miranda House. Miranda is the name of Prospero's beautiful daughter – the main lady character of William Shakespeare's classic, The Tempest. In the play Miranda is naïve and innocent yet she stands up for herself to achieve her goals in life. Perhaps Gwyer sought such virtues in the grooming of young women of India and was thus inspired for selecting this nomenclature for the college. The name for the college endures.

Designed by the renowned architect Walter George, the warm red brick of Miranda and its cool and spacious corridors resembled the architecture of the convent school where I had studied, so it felt most reassuring. But here ended any similarity between my school and college. The two were the most dissimilar institutions I had ever seen or attended. Unlike my clean, quiet and manicured convent, Miranda of those days in 1976 was a place where bathrooms leaked, classrooms were untidy, the canteen was rusty and the gardens more a wilderness of sorts. The principal was Dr A.C. Janakiamma, and my family would use the suffix of her name "amma" in the most humorous of ways! In any case Amma was fond of me as I had won the competition for designing the Silver Jubilee shield for the college.

Moreover, I participated in the larger life of the college, won essay writing and poster competitions, and participated in debates and cultural programmes. I was elected Secretary of the Association of Geography Students, a small-time post that I took rather seriously, setting down a bridge for many brewing issues which ranged from the availability of books in the library to maps in the department and the squabbles over those boyfriends who enjoyed knocking at the famous back gate of Miranda House!

When I joined Miranda as a student, the country was in the grip of a State of Emergency declared by the country's prime minister, Indira Gandhi. Emergency was imposed on 25 June 1975 and a large number of teachers had been put under house arrest. All public meetings and gatherings were banned. The Miranda House faculty, like many other teachers of that generation, raised a strong voice against the suspension of democracy and curbing of the freedom of speech. As a result, there was little festivity and few cultural activities at the college in the year I joined. The dark phase of Emergency came to an end on 21 March 1977. The first non-congress government swept to power and, following the defeat of the congress at the general elections, Morarji Desai replaced Indira Gandhi as prime minister. The Mandal Commission was set up to identify socially or educationally backward classes, to consider the question of reservations for people to redress caste discrimination. The turnover of political events was swift and by the end of that year, Indira Gandhi was re-elected as prime minister of India.

It is therefore at Miranda that I experienced my first close contact with India; the merits of its democracy and the specificities of its disparity and diversity. It is here that I realized how privileged my schooling and ambience had been. Among my batch of 30 odd students there were those who stood in long queues to seek monetary concession for the monthly fees of Rs 30. There were those who wrote entire lectures in English and translated it into Hindi to learn for exams, and those who would rarely eat in the subsidized canteen as food was an expense they could ill afford. There were students from different parts of India, with a distinct group from West Bengal and Punjab. I had never witnessed so many cultures, nor encountered such a latitude of economic classes. A new deeper sense of India began to ignite in me. I was humbled to be there.

Whatever be their origin or economic status, a common feature of discussion among girls was their concern of what to do and what to become after graduation. I was quick to make friends and struck a deep affinity with a pretty Kashmiri girl who later became the reason for my choice of Kashmir Valley as my field of research for a doctoral degree. Later, I had the opportunity to meet my friend at her home in Melbourne when in 2010, Vice Chancellor, Deepak Pental, University of Delhi (2005–2010), nominated me as visiting research fellow, New South Wales University, Australia. We spent hours reminiscing the memorable days at Miranda House, those university bus specials which ferried us cheap and fast, the hit-the-butt-and-grind-it-well Miranda walk, monsoon days filled with *chai* and *samosa* and the different fashions which changed it all.

During my undergraduate years, R.C. Mehrotra was vice chancellor at the University of Delhi, (1974–1979); for a student at my level, he was an authority one rarely got a chance even to see or meet. Who mattered most was the faculty of geography at the college and they were an interesting bunch. I enjoyed the precision with which Punam Behari taught statistical methods and map-making, the glamour which Rashmi De Roy brought to the class of human geography, and the enthusiasm with which Nirmal Kandhari taught us geomorphology and trained us in the practicals. The classes on India and economic geography by Surinder Sahi were long and at times difficult and the slightly scatter-brained Daleep Chaudhuri left me exasperated. I cannot, however, forget the dainty Mamta Mayee Sharma who taught us climatology. She taught this paper in such a difficult tone and tenor that I had to work harder for its understanding. There was a dividend, of course. I scored the highest in climatology. But it was with Nirmal Kandhari that I kept a lasting relationship as a tribute to her commitment to teach with immense passion and effect.

The classrooms were small and I recall tracing tables designed with rudimentary tube lights below a glass surface. These were kept in different corners of the classroom to draw the maps. Interpreting toposheets and the synoptic weather charts were weekly routines and I simply found joy in the pattern of wind flow which crossed those isotherms and isobars. Miranda House also had the standard wet and dry bulb maximum and minimum thermometer in a Stevenson Screen alongside a barometer and rain gauge. For me, this weather lab seemed so scientific and I would visit it often.

Those were the days of surveying with instruments like the plane table, chain and tape and the prismatic compass. With these weighty instruments we would locate the flower beds, bottle palms and pavements in the quadrangle of the hostel of Miranda House to scaled perfection. I was learning how a map is made.

The undergraduate classes taught me how rivers run and erode, monsoons originate and burst, cities grow and expand, how regions, regionalization and regionalism are different from each other and so much more. I took the BA programme most seriously and became absorbed in the fundamentals of every course. Books by V.C. Finch, G.T. Trewartha, W.D. Thornbury, and the father and son team of A. Strahler and A. Strahler, were among my favourites.

It was at Miranda that I learnt the role of the legend, the proportionate circle and the choice between the symbol, shading vs using lines as a method of representing data. All this proved very useful when I began to teach a course on cartography and more so when I set to design a variety of maps for my books and exhibitions. A map is clearly a powerful intellectual tool. Sometimes I wonder how one would go about explaining location, direction and patterns without this tool that is so taken for granted.

A special delight was the field visit to Chamba, an erstwhile princely capital on the banks of the Ravi river in Himachal Pradesh. I can never forget the moment at Khajjiar, when my teacher pointed out that I was standing in a cirque which had a small tarn! It was perhaps the first time that I came to realize how important it is to take the class to the field. I worked hard at my field report and the department gave me a certificate which read, "...presented a good study report on the Dalhousie – Chamba." Somewhere I began to see that I had the makings of a geographer. I scored a high second division in my under-graduation and applied for the postgraduate programme at the Department of Geography, University of Delhi with a determination to learn deeper and do better.

I enrolled as a postgraduate student in 1979, when I was hardly 20 years of age. I completed my formal higher education with a Master of Geography followed by a Master of Philosophy in Geography and a doctorate in 1989, at the age of 30. A clear decade. And yet today it seems like the fastest period of my life where I was on a high speed skate of learning, free from any other thought and commitment. No place proved better suited for me than the Department of Geography at the Delhi School of Economics.

As a student at Delhi School of Economics

Maurice Gwyer, the longest serving vice chancellor, University of Delhi, appointed Dr V.K.R.V. Rao as the first Professor of Economics in 1942. Rao announced his plan to establish the Delhi School of Economics (DSchool) on the lines of the London School of Economics and left no stone unturned to realize his grand project. The DSchool Society, with Prime Minister Jawaharlal Nehru as its president, was formally registered in 1949, a year considered as the birth of DSchool.

Rao wanted DSchool to have its own building and campus. He secured from the university a nine-acre plot of land which now houses the departments of Economics, Sociology and Geography. The new institution attracted the best of minds in economics. Mention can be made here of Amartya Sen, Jagdish Bhagwati, Sukhomoy Chakravorthy and Manmohan Singh, among others. Their seminal contribution catapulted DSchool to international eminence.

It is within this habitat that a department of geography was created in 1959. The initial run of its classes was held in the main building of the DSchool but when a new teaching block was inaugurated in 1960, its first floor was allocated to geography. Subsequently some rooms on the ground floor were also given over. It is on these premises that the department of geography continues to date. The title, Department of Human Geography, was the choice of V. K. R.V Rao, vice chancellor, University of Delhi.

The official records state that

> ...we have named this department, as the Department of Human Geography for a specific reason: when students of economics study problems related to agricultural economics, they do not know anything about the natural conditions of the region; when we had to select certain clusters

of regions for the purpose of our study, we found that we were right at the heart of the problem which was of a geographical character. We thought we needed a Department of Human Geography for the simple reason that we wanted to emphasize that a particular part of geography is closely allied to economics, political science and also agricultural economics

The above was the explanation given by then Director of the DSchool, B.N. Ganguli, in the inaugural year of the Department of Geography.

A factored reason was the Department of Sociology, which was also inaugurated in 1959. V.K.R.V. Rao was keen to induct M.N. Srinivas into the Department of Sociology. Srinivas's wife, Rukmani was a geographer and she was also to be accommodated. To serve the dual purpose, the Department of Geography was established in the same year as a constituent of DSchool.

To make the department more holistic in its teaching and research, which embraced both physical and human geography, the prefix human was dropped from its title in 1976, three years before I joined the department. My transition from undergraduate to the post-graduation programme was crucial. The general atmosphere at DSchool was sombre. DSchool was neither a part of the strikes of the Delhi University Teachers Associations and nor were students the members of the Delhi University Students Union. The prime focus of teachers at DSchool was to impart high quality education through teaching and research.

DSchool provided an ambience which I instantly fell in love with, the imposing red brick main building, sprawling gardens and the many species of trees scattered randomly with no order or design were scenically most appealing. Ratan Tata Library was the hub which stayed open until late into the night. There was no photocopying facility and we spent hours making notes, at times using carbon paper to make a copy for our friends. Those were the days of chalk and board teaching and rudimentary rotary pens whose ink was either drying or spilling as we drew maps placing our tracing paper atop the graph paper.

My school and undergraduate college had been all-girl institutions and it is here at DSchool that I experienced my first taste of co-education. Interacting with the boys, I began to shed my characteristic shyness and some quaint convent mannerisms.

In 1979, when I enrolled for post-graduation, there was word around that stalwarts of the Department of Economics had all but left. This apparently was true. The departure of Jagdish Bhagwati, Manmohan Singh, Amartya Sen and K.N. Raj from DSchool to take up positions elsewhere was referred to as an exodus.

Attrition had also happened in the Department of Geography. S.S. Bhatia, Amrit Lal, and Ranjit Tirtha had resigned in the mid-1960s while the star, Professor V.L.S. Prakasa Rao left the Department of Geography in 1973. I have never met or seen any of them. In spite of these departures I felt that when I enrolled as a student it was a time of the rising sun for the Department of Geography. This is because the faculty which joined were dedicated and committed to serving the institution until their retirement. Moreover, they came from different backgrounds and with varying interests; some had acquired their higher degree from American or British universities while others had joined straight after completing their degrees at Indian universities.

Both R. Ramachandran and S.G. Burman had taken their doctorates at Clark University, in the US, C.P. Singh had obtained his degree from the School of Oriental and African Studies, University of London and B. Thakur had studied at the University of Waterloo, Canada. On the panel were also S.K. Pal from Jawaharlal Nehru University, New Delhi and N. Mohammad from Aligarh Muslim University, Aligarh. While their origins added variety, their areas of specialization provided for highly diversified learning. There were courses which addressed issues of resources, biogeography and environmental studies, agricultural, industrial and political geography. There were those which dealt with contemporary social relevance taking geography closer to urban and regional planning and sustainability. For me, this variety continued to add colour and flavour to the personality of the discipline and I enjoyed all the courses equally. The favourites were professors and not courses. This meant that I studied each course with equal interest, intent and intensity.

What I enjoyed most were the lectures on the theory of location. These were spread across three different courses of the syllabus, agricultural, industrial and urban geography. The theory of concentric zones of agricultural land utilization around an isolated city put forth by Von Thunen was taught by N. Mohammad. He drew the diagram and explained the model carefully. Alfred Weber's comprehensive theory of the location of manufacturing was explained with ready details by C.P. Singh, while R. Ramachandran taught Christaller's central place theory that dealt with the size, number and distribution of settlements. I was most impressed with how Christaller discarded Von Thunen's circular complimentary areas and replaced them with hexagons. I often wondered if location theories represent reality?

B. Thakur and N. Mohammad were methodical in their lectures. Both had a neat script and wrote detailed outlines on the board; they both spoke at a writing pace and at times it seemed like a virtual dictation. N. Mohammad was incredibly meticulous when he taught cartography but he was rather unimaginative. I had to take a repeat in his course because he insisted that making a map is a science and I appeared to treat it like an art! I was keen on maps not just as a tool but also as an aesthetic experience. In later years I learnt how to blend the science with the art. At the other extreme was S.K. Pal who taught us statistics and geomorphology in what seemed to me a most confused and mixed-up way, but he was friendly and laughed easily and seemed so absorbed in the numbers that I often wondered if he was even aware of us students being around.

The intellectual and social highlights of that time were many but nothing compared to the lectures by R. Ramachandran throughout the year. His lectures were well rehearsed, regularly scheduled for two hours without a break, and what the small class of 22 students was offered each time was an introduction to the concept, followed by a Western interpretation, and then a case study. I looked forward most to the last half-an-hour or so of his lecture where he would provide an alternative Indian perspective, based mostly on his own concepts and research. Whether discussing Christaller's central place theory in an attempt to explain the spatial arrangement of size and spacing of settlements in Uttar Pradesh, or illustrating Torsten Hägerstrand's innovation diffusion model as a spatial process when applied to the spread of tube-well irrigation in Tamil Nadu, Ramachandran took an overview with the clarity that comes from conviction. While the overall curriculum was a bastion of the West, it was through this mode that his interpretations paved the way for the Indianization or Indigenization of the inquiry. His background in mathematics and hands-on work on the first generation of computers gave him the skill to crunch numbers and discard them with equal ease. He personified many virtues. In one of many of our conversations, he shared, "I never wanted to be or liked to be a leader and the leadership game among academics did not interest me." I think there are different ways to lead and he was inspiration.

I had read the neat handwritten manuscript of his book on urbanization which took shape during his lectures to us in class. In structure and style I thought it was heavily influenced by Raymond E. Murphy's *The American City: Urban Geography*. After all Ramachandran had taken his doctorate at the University of Clark. Ramachandran's script took an expansive view of the evolution and patterns of urbanization in India, followed by chapters on the internal structure of the Indian city and then on to the problems and policies for urban places in India. His book was published in 1989, the year I submitted my doctorate. I was sad to learn that Oxford publishers had managed to push him into breaking it into two parts and published the first half under the title, "Urbanisation and Urban Systems in India," and when the turn for the second part came Ramachandran decided against its publication. I still regret not photocopying the entire draft, as I perhaps could have prevailed on him to send the remainder to press. It is an academic loss, especially when one knows that the half which was published has maintained its relevance and eminence for the last three decades. I recall here my role in ensuring a publication of its Hindi version. The anecdote is worth telling.

On a cold winter morning, Dinesh Singh, vice chancellor, University of Delhi (2010–2015) walked unannounced, into my class in the department. This was a feature common to his style of administration and concern for academics. On his interaction with the students there emerged a need for texts in Hindi. The vice chancellor asked me to help fulfil this demand raised by the students. I opted to pursue the case for a Hindi version of Ramachandran's book. A friend who is the proprietor of Vitasta Publishing

House was ready to take the initiative. The latter was released at a function when I was chairperson of the Department of Geography in 2011. To this day, I continue to enjoy a warm academic amity with Ramachandran.

If the high in R. Ramachandran's classes whet my appetite to develop a birds-eye view about the geography of India, S.G. Burman's classes on ecology and the environment were a completely different genre. She lectured from the ground, and her forte was the Himalayas. I remember her coming to our class with a tray of slides and an armful of photographs to augment her lectures. Her visuals based on travel and field experiments provided a window, especially of the Himalayas, and were all the more meaningful because of the personal stories that accompanied each frame shown on screen. She was highly possessive and had devised her own method of coding and classifying her slides and photographs. When she died no one could archive these, but for her it was a life shared with students and that is what matters.

Burman had a rather relaxed style and she inspired by emphasizing the need to back ideas with facts and not with the fluff of jargon! Here Burman marked a distinction from Ramachandran. Burman came with narratives, case studies and descriptive details while Ramachandran was profuse with concepts, models and theories. Both were conscientious and devoted to scholarship. While I set out to do my doctorate with Burman, it was my learning from Ramachandran which shaped my writings of later years.

The post-graduation programme flew in a blink and this time round I scored a first division and was a front ranker among my batch. Since I was enjoying the world of books, enrolling for a research degree seemed a natural decision. By this time, many members of our family had emigrated and settled in America and England. I could well have considered taking my doctorate from a foreign university. I recall having an extended discussion with R. Ramachandran in this regard and he sincerely advised that a PhD programme is more about self-learning and if I wanted to spend my life in India then a worthy work could well be accomplished here. It was a decision I have never regretted.

Research apprentice

Spurred by my supervisor's enthusiasm and love for fieldwork in the Himalayas I chose Kashmir Valley as my field; an added incentive was my Kashmiri friend from the Miranda days. She accompanied me on many field visits as I lived on houseboats, rowed the *shikara* on the River Jhelum, surveyed hotels in the hill resort of Pahalgam, visited looms which spun pashminas and camped amidst the apple orchards. My most memorable hours were when I cycled, with the wind in my hair, absorbing the breathtaking beauty around the Dal and Wular lakes. In my heart I carried sorrow, as I came to discover that both the lakes were being heavily polluted and soon would shrink and die.

I submitted a doctorate entitled *Ecological Implications of Changing Land Use in the Kashmir Valley*. The research allowed me the opportunity to chart mostly my own course. I strongly believe that a PhD has to be mainly self-propelled with a little bit of supervision. I came to accept the truth of what someone once said: the best way to learn is to learn how to learn.

Typing endless drafts on my Olivetti and hand drawing maps was a tedious feat and I was glad when finally the viva voce certificate read, "She has been found successful for the award of the Ph.D. Degree of the University of Delhi." I was conferred the degree in 1989 by Vice Chancellor Moonis Raza, University of Delhi (1985–2000). Little did I foresee that a decade later I would be writing a biographical note on him. Among other things, Moonis had been the president of the National Association of Geographers, India, the largest body of geographers in India.

I am tempted to describe the ten years of 1979–1989 as the cocoon phase of my life. A time when I was completely wrapped up trying to prove the thesis of the ecological ruin of the Kashmir Valley. It was a passion so deep that it turned into an obsession. I used to get nightmares about the shrinking of Dal Lake and would let no one eat those Red Delicious apples as I knew first-hand that they had been drenched in pesticides many times before they made their way from the orchards in Kashmir Valley to Delhi! I now can sense how tiresome I must have sounded to my family as I repeatedly described the Bakerwals and the Gujjars who made their seasonal transhumance, and how strange my friends might have felt when I endlessly moaned about the springs that were dying within the premises of Shalimar and Nishat, those legendary Mughal Gardens.

I made several trips to the valley and it became my second home. I was deeply disturbed when insurgency blocked my visits after 1989. Thankfully by then I had obtained my doctorate.

While I was engrossed in the facts, field notes and books on geography life, had flowed past. My sister was married in the UK with two children. My brother married a Swedish girl and they had two boys and also moved away to the UK. Both my paternal and maternal grandparents, all noble souls, were no more in this world. My mother ventured to take a diploma in art and design and set up a garment business; my father was wrapping up his business of manufacturing watches and timepieces. These were years when India began experimenting with the idea of Export Promotion Zones (EPZ). My father took this opportunity to get a licence to export jewellery and set up his manufacturing in Jhandewalan, an industrial estate close to our home. When the government of India enacted the Special Economic Zones Act in 2005, my father upgraded his establishment to this status.

At the national level, during the years spanning until 1990s, the prime minister of India, Indira Gandhi, was assassinated in 1984, the Bhopal gas tragedy had spewed its poisonous gas, killing an estimated 25,000 people in the same year, and the Environment Protection Act was put in place in 1986. Within all these changes, I stepped out of the mould of formal learning to seek a space to teach geography. The University of Delhi was to be my destiny and destination.

Seeking space course to course, college to college

When Maurice Gwyer laid the foundation of the University of Delhi in 1922, he modelled his vision on the lines of the University of London, which consisted of both university departments that offered postgraduate teaching and research, and colleges which offered an undergraduate programme. Such an arrangement of departments and colleges is not a pattern followed by all universities in the country. For example, Jawaharlal Nehru University and Jamia Milia Islamia University are both located in Delhi and although both offer postgraduate teaching and research neither has constituent colleges for undergraduate teaching.

When I put up my first job application in 1987, the University of Delhi had 65 colleges but geography was offered in only six. I later learnt that such a languishing condition of the discipline was not just a feature of the University of Delhi but was typical across many centres of higher education in India. I was curious to know why this was the case but I postponed analyzing the reasons for such a state of affairs until I secured a space to teach geography. The search was more difficult than I could ever have foreseen.

During 1987 to 1997, I forwarded 30 applications, was interviewed 20 times and took whatever tenures came my way. I taught at an all-girls college and a co-educational one too. I taught in colleges within campus and those located outside the main campus. I agreed to take teaching jobs with a daily wage, a part-time salary, full-time salary, salary with holidays and even one where holidays were not paid! This meant my earnings were highly irregular and uncertain. The only grace was that I had a class to teach each day of these long years. This kept my skates rolling through this bumpy time.

Timetables pulled from colleges will confirm the large latitude of courses I taught. This was not so much out of choice but more by circumstance. As an undergraduate teacher I was mostly on a temporary tenure which meant that I had little choice in the courses that were handed out to me to teach. Since I had to hunt for a job every year and sometimes within six months, this meant bits and pieces of many courses came my way. When I joined DSchool most faculty had been my teachers I again had to make do with what was doled out to me. It was an ordeal to develop teaching material anew, and the only solace I now draw is that this diversity has given me the ability to pull together many strings of different courses for effective synthesis and analysis.

Teaching came to me almost naturally. The convent I attended had given me a command of language and some of the best and not so worse teachers had taught me the merit of combining serious lectures with storytelling and humour. It was sheer fun being with these fresh earnest minds straight out of school; it seemed both they and me were together searching our way through life.

Teaching in colleges had other virtues. During this pastoral movement across different colleges I made many good friends who, to date, endure. At times, the situations turned turbulent. A serious blow was an interview at a women's college. I had applied with a doctorate and first division but another candidate with a lesser qualification was selected. I was unhappy. But to my surprise nearly 30 years later, I was appointed as a member of the governing body of the same college. When I chaired my first governing body meeting, many thoughts crossed my mind. I realized that not all storms come to disrupt our life; some come to clear a path.

I had built a reputation as a good teacher and yet was not able to obtain a secure job in spite of several interviews; perhaps it is a truism that it takes talent to recognize talent. It was only when Leela Seth, the first woman judge of the Delhi High Court, and Kiran Datar, principal of Miranda House – honoured with the Mahila Shiromani Award (1994) – came together as members of the selection committee, that I finally secured a permanent tenure as assistant professor in Miranda House in 1995. It had taken me 10 years, to get a permanent space to teach geography! Sustaining my emotional agility was getting a bit difficult.

Under the tutelage of Kiran Datar, Miranda House carried a sheen and shine in sharp contrast to the lax and unkempt ways of my student days. The college was all set to celebrate its golden jubilee in 1997. There was a lot of excitement and daily deliberations on events to be organized. I was nominated member of the Golden Jubilee Committee and while I wanted to give back to my alma mater, my soul now longed for not just a place but a space of my own. This was because I was teaching out of the textbook and not saying anything that was intellectually mine. I longed for a reflective space where I could develop the many geographical ideas brewing in my mind.

Just when I was yearning along these lines, the Department of Geography, DSchool advertised a post for assistant professor. It had taken 19 years for the department to announce this one single vacancy and guess what? I was selected. I was actually selected! Those were days when the university dispatched telegrams. On a Monday morning 14 April 1997, arrived the clipped message .... *Anu Kapur selected lecturer, Department of Geography, University of Delhi. To join at the earliest.* Though it was not the earliest for me as I was nearing 38 years of age, yet better late than never, and I cried like a baby.

I remember the University of Delhi's vice chancellor, V.R. Mehta (1995–2000), had asked during the interview "Which foreign university have you studied at, for your articulation of ideas in English is crisp and carries great clarity!" I was filled with pride to reply: "University of Delhi Sir!" As life would unfold, I was to get the chance to thank V.R. Mehta for playing a role in my selection only when his son, Pratap Bhanu Mehta and I were among the ten social scientists bestowed with the Amartya Sen Award for Distinguished Social Scientist by the Indian Council of Social Science Research, New Delhi. This meeting was to wait a good 15 years.

Life is dense. Ten years of post to post, interview to interview, course to course and college to college had told me and shown me that in India, population, pollution and politics is dense. To make one's own space in academics it helps to be guided by an inner sense.

Many waters had flowed since the year I enrolled as a postgraduate student in 1979 and when I came to join as faculty in 1997. My doctorate supervisor Burman had died and Ramachandran had retired in 1996. To add to the loss was the demise of C.P. Singh in 2000. I missed their towering presence.

Mother Teresa's Missionaries of Charity runs an orphanage at the Civil Lines, an address just ten minutes' drive away from my DSchool. I frequent this place to play and be with the children. On one of my visits, I met Mother in the chapel. From close watch she carried a deft firmness around her mouth and the most compassionate set of eyes I could ever have found. I realized that it took both resolute firmness and deep compassion to make the blue saree border institution such a success. In 1979, the year I enrolled as a student at the DSchool, Mother Teresa was awarded the Nobel Peace Prize for her humanitarian work with the poor. She died in 1997, the year I joined as faculty at the DSchool. For reasons unknown I mourned her loss as personal. Today in 2020 I have crossed my 60th birthday. In these years what has geography given to me, what has geography been for me, and how did I engage with geography?

What geography gave me

Geography enabled me to climb the professional ladder from assistant professor to associate professor and further to professor. My financial earnings inched up very slowly in the beginning staying stagnant for years, and yet I managed to climb to a high rung in this profession. The graph resembles the J-curved shape of India's population growth. Its explanation rests with the recommendations of pay commissions which the government of India sets up at regular intervals. The 6th Pay Commission of 2006 gifted a bonanza. Gratitude remains.

A major time of my life in the department has centred around teaching. We were short staffed for years and this meant having to teach many courses that were compulsory on the syllabus. As for students they have been the backbone of life at DSchool. I like the ambience of students wandering around in corridors, hanging out in groups on the lawns, the general activity of students walking in and out of classrooms, and the feel of the library that augurs high purpose and noble goals. I find teaching exciting and could never take my class for granted. I am known for those longish classes that spill beyond two hours where PowerPoint presentations are few and along with serious explanation the lecture is peppered with humour. I have received letters and emails filled with the most endearing feedback along with suggestions to punctuate the long class with breathers and breaks, speak a bit more slowly, how I may recommend more readings and so on and so forth. The most memorable compliments I have received are those from students who wrote that they felt equipped with altogether new perspectives on themes familiar to them, saying that for them a new kind of geography had been ignited and that time just flew while listening to me! With nearly 90 students in the class, keeping a 180-degree alertness in my lectures scheduled on an early Monday morning has always been an encounter of sorts.

Generous with my teaching time, I was perhaps not so liberal with marks. I wanted to separate the serious from the flippant and also did not want to miss out on those who worked hard to learn. This meant the graph of my marks carried a double hump, one for the best and the second for those who showed all characteristics of making sincere effort. In the saddle fell those who got low grades, they do crib and I am still to learn what to do with those who are casual and careless.

I know that one is dealing with students in their formative years when they are crystallizing their view of the world and life. Moreover, many of these students are going to wind up in positions of some kind of responsibility and influence. In my own mind, I was ever keen to invoke a sense where they could realize that there are many different ways to interpret geographical reality. Changing their perceptions was not always easy. To me their unlearning was a greater challenge than the learning.

I taught with passion and compassion but I confess I did not keep in touch with past students. I also must add that I have not been able to build a coterie of loyal disciples who would take forward my ideas and approach to research and teaching. Perhaps since I valued my own freedom so fiercely I also did not want them to be tied to any apron strings.

I taught best what I had researched most. Much as I would have wanted to knit my research into teaching this was not easy, simply because what I researched had to wait before the syllabus was revised. The latter has become a slow process. We have been teaching the same syllabus for 13 years and it was only when the National Academic Accreditation Committee visit was scheduled that we were compelled to revise our syllabus in 2018. It is only now that I have been able to incorporate two entirely new courses in the department. One is titled *Geography in India* and the other *Geography of India*. The former traces the history and characteristics of the discipline. In the case of the latter, though the title seems hackneyed, the contents cover new themes like vulnerability, disasters, vitality, place names and place goods, and also modules on displacement and disability in India. I am afraid that not keeping the syllabus and reading abreast with the times would eventually implicate the status of the discipline in the country severely.

At the department I had the opportunity to engage with the postgraduate and research students. I could supervise dissertations on themes as varied as disasters, environment, pollution, traffic, crime and ethnic violence. There were studies on old people, those disabled, rendered orphaned or without shelter and those left without work. Pilgrims, tourists, refugees and migrant workers were not excluded either. The field of students acquainted me with the ground reality of problems and places in virtually every state and corner of India.

The enormous diversity within the subject of geography gave me the opportunity to explore and harness a wide variety of data. To estimate landslides on the Banihal Pass – a road which connects Delhi to Jammu to Srinagar – I have searched the headlines of the local newspaper *Kashmir Times* for a period of 10 years. To calculate visitor numbers to the Mughal Gardens in the Kashmir Valley, records of the sale of entry tickets were checked. The first-hand inquiry reports at the police station were noted when plotting the number of crimes on campus at the University of Delhi. I visited the National Philatelic Museum at the Department of Post, Delhi, to handpick if they had used place goods as images on postal stamps. I mailed endless Right for Information requests to the Geographical Indications Office at Chennai when I needed details on place goods and did the same to the office of the Delhi Metro Rail Corporation, when I needed passenger movements at different stations on the metro in Delhi. The websites of the High Court were browsed to extract legal cases of fire disaster in India and the same source of data was found rich when I was seeking infringements against place goods in India. The debates among parliamentarians in the houses of Rajya Sabha and Lok Sabha were rich on narrative about how administrative states and union territories are named. The Working Plan Documents of forests catalogued in the Forest Research Institute, Dehradun, the climatic data from the Meteorological Department, and details about population characteristics from the Census of India, among a host of other sources were never to be forgotten when building a spatial inquiry.

It appears that I wanted to use all sorts of data to uncover spatiality. Balancing between theory and empirical elaboration has been fun. I always seem to be in a hunter and gatherer mode; nothing was easy but likewise nothing was impossible. It is always hard for me to understand how data could stop one from exploring a fresh field of study. I agree it consumed a lot of time and was painstakingly slow but the analysis was the exciting part where like a jigsaw puzzle suddenly all that was gathered would begin to make sense. It is this elation which would pull me into another round of search and research. Geography gave me the scope for all this and more.

The subject showered other unexpected joys. I had heard about Professor R.N. Dubey and his seminal contribution to the Department of Geography, University of Allahabad. I also knew that his birth centenary, 15 November 1996, coincided with the golden jubilee of that department. On this momentous occasion a generous endowment from the Dubey family led to the formation of the R.N. Dubey Foundation. What I did not know was that I would one day be taking home an award with the name R.N. Dubey and here is how this became possible.

In 2000 K.V. Sundaram, an enthusiastic geographer, had spearheaded the formation of a consortium of Earth sciences which he fondly named the Bhoovigyan Vikas Foundation. I know from personal interaction that Sundaram's generous family had supported this venture, both materially and intellectually. It takes sacrifice to create good. When two foundations hold hands much can be achieved. The R.N. Dubey Foundation made a generous contribution to the Bhoovigyan Vikas Foundation. This made possible the initiation of a set of awards – one for senior geographers above age 45 and the other for middle level geographers below this cut-off. I was one of the four geographers in the country who were awarded the Bhoovigyan Vikas and R.N. Dubey Leadership Middle Level Award in 2001. It carried a well-versed citation along with a token cheque. While I was truly elated, a bigger surprise was yet to arrive.

In 2012 the Human Resource Development Ministry announced an award for ten top social scientists in India in the name of Nobel laureate Amartya Sen. I had heard Sen at an inaugural lecture for DSchool's golden jubilee on 14 November 1999. Little did I know then that I would one day carry his name home in an award that would be given to me by Sen in person.

> ....Anu has made a transformational impact on Indian Geography by way of generating fresh discourse and initiating a debate on the state of the discipline in India. Her eminence spreads itself in the wider arena of social sciences, with a firm footing in the theory and methodology of Geography.

This is the highest and the most prestigious Social Science Award in India and carried a citation and prize of Rs 10 lakhs. I hankered after none of this yet it felt good. Thank you geography!

Chairperson at my Alma Mater: When I took over the reins as chairperson of the department in 2011, I was enthused to use the opportunity to give back to my alma mater. A new office was raised, classroom size enlarged and faculty rooms added. This was accompanied by strengthening of infrastructure. The computer lab was augmented with the latest software and audio visual facilities. There was something more to be done. Three thousand maps stored in the corridors of the department were shifted to the Ratan Tata Library for access by social scientists at large. My most favourite was landscaping the garden and I have a compelling need to describe it:

> In an oblong shaped patch at the entrance gate of the Department of Geography grows a tall *bargad* or banyan tree. This tree towers magnificently and has wide reaching roots. I have not been able to figure out its age and the eldest of *malis* or gardeners say it has just forever been here. The tree is revered in India and this one carries such a shield of protection and draws care that when I took administrative responsibility of the Department I raised the entire oblong patch within the vicinity of the tree by a good four feet or so. This allowed a scope to plant other tropical species over the raised mound. With the help of a local crane large boulders were carried from the Aravalli ridge and set

at the base of the tree. I sculpted the word Geography onto one of the rocks. To complete the picture, I placed an upturned round earthen pitcher into the central hollow of the trunk of the tree. It was meant to symbolize the earth. It pleases me when students and visitors use this picturesque landscape as a backdrop for their memorable *selfies* and group photographs. If *bargad* was a sacred tree at the entrance of the Department, the window of my study opened onto the beautiful *shisham*. Under its shade, I spent many winters reading and writing. I feel a mystic bonding with nature.

I tried to inculcate a culture of autonomy for every faculty and used the time to invite eminent scholars across disciplines for lectures to the students. I also used the opportunity to create a special function for the farewell of three colleagues from my department, who had retired before my term as chairperson but without the grace of this ceremony. None can deny that politics in departments can turn so rife that we forget even basic courtesies and human sensitivities.

As an obligation to my university, I served on a few committees within the University of Delhi and at the National Council of Education Research, Delhi; the School of Planning and Architecture, Delhi; and governing bodies of prestigious colleges such as Lady Shri Ram College for Women, Kamla Nehru College for Women, and the Delhi College of Art. But administration was definitely not much to my liking, I rued the time it took away from teaching and writing. Currently I feel relieved that young faculty have joined and taken over many administrative chores.

I must confess that I have pursued my own interests with relatively little attention to what was exciting many of my colleagues. This was not so much out of indifference to them but more to do with my mental preoccupations. In the acknowledgements of my book *Mapping Place Names in India* (Routledge 2018), I wrote: "My colleagues tell me little, ask me little, meet me little. They let me be. That is not little." I too am the same with them. I do not know how this space evolved but what I do know is that it kept my long hours at DSchool tranquil and joyful.

At DSchool I tried not to be a challenge. I refrained from being a member of any interest group, but when a difficult colleague was always breathing down my neck, I had the courage to take the issue to the university administration. Although most of my colleagues in the department supported my case, as soon as the administration at the helm of affairs changed, some among these same colleagues who had signed in support of me made a U-turn. What I learnt is that, since our own colleagues bend to appease those in power, the system fosters harassment.

Yet my pride in being faculty at DSchool never diminished. When I joined as a fellow at the Indian Institute of Advanced Studies (IIAS) in Shimla during the period 2003–2006, and as senior fellow at Nehru Memorial Museum Library (NMML), New Delhi in 2012/2013, my effort was to complete the assignment well before time so that I could return to my workplace. It is not that the two advanced study centres were places of any lesser grandeur. I continue to carry the "iias" as part of my email id.

At Centres of Advanced Study: Perched amidst a scenic setting of what is called the Summer Hill, the IIAS, Shimla, was designed by Henry Irwin, who used the grey stone facade and the Burmese teak interior to create a breathtaking structure. My study opened onto a patio which looked out at the snowcaps of the Great Himalayas. The building was originally built as a home for Lord Dufferin, viceroy of India from 1884 to 1888 and was called the Viceregal Lodge. It hosted all the subsequent viceroys of India. Some of the most historic decisions about India, including Partition, had been taken within its premises. After independence, it was renamed as Rashtrapati Niwas, residence for the president of India. In 1965, the president of India, Dr S. Radhakrishnan, decided to change the occupancy of the place from a president's house to a *niwas* (resort) for social scientists. Here is the first-hand explanation by him on the occasion of the inauguration function: " – When I took up, my office as President, I looked into the statistics of visits to Shimla and was informed that in a period of 15 years we [had] spent about 120 days here, less than 10 days a year, ... so after consulting the Prime Minister, Jawahar Lal Nehru I said, it should be devoted a more useful purpose than merely the pastime of the President."

I felt most fortunate when, after my tenure as fellow in the president's house, I was selected senior fellow in the former prime minister's home now called Nehru Memorial Museum and Library (NMML), New Delhi. Set within a 30-acre campus, its stately building was known as Flagstaff House and was used by the British as the residence of the Commander-in-Chief. This place became the residence of Jawaharlal Nehru for the years 1947–1964. After his death in 1964, it was converted into a centre for advanced studies and a venue for visitors to sightsee. Unlike IIAS, NMML does not offer a residential facility. This suited me perfectly; I had the best of both worlds: a perfect place to research, and my own perfect home to stay. The value of the centres of advanced studies was enormous. Sitting in my study uninterrupted without the bell for the timetable or procedural meetings and allowed to write on a subject close to heart, and the wondering and wandering in their magnificent gardens was to me a precious time and place. I strongly believe that freedom in a place is important to allow space for thoughts.

The centres gave me an opportunity to interact with social scientists from other disciplines, and I benefitted greatly from scholars like Pabitra Roy's exceptional clarity in discussing both Western and Eastern philosophy; renowned Indologist Bettina Baumer's discourses on Kashmir Saivism; and Navjyoti Singh's endless debates on the *nyaya* (justice) school and its treatise on epistemology. I can never ever forget Karuna Goswami, a historian and her husband, B.N. Goswami, the famous art historian. I attended all his seminars and was so inspired that, in 2008, I enrolled for a course on art appreciation at the National Museum of India, Delhi.

Both the Indian Institute of Advanced Studies, Shimla and the Nehru Memorial Museum and Library, New Delhi are leading centres for learning and research in social sciences and humanities in India. Each year the number of selections at these centres is around 25 academics from across India. I was the first geographer in India to be selected fellow and senior fellow, respectively, at both these two centres. This gave me a sense of pride, yet it also filled me with remorse as to why geographers before me had not knocked on these doors. At the advanced study centres my genuine sense of inquiry, the belief that things of the mind are worth pursuing and a concern for exactness and precision was most dominant. I remember a scholar at the IIAS once remarked that "I was most catholic about my research." My convent values seemed to endure. Both were magical places to write a book. I quickly fell into a rhythm of reading, thinking and writing, yet submitting my manuscripts well ahead in time I headed back to DSchool, my own space.

Engaging with geography

In contrast to Miranda House the Department of Geography, DSchool was a totally different place. Miranda carried a buzz of festivals, many societies and clubs, streams of students walking across corridors, the teachers belonging to different disciplines. In contrast, the ambience of the department marked a departure. Its size was small, it had only two lecture halls, one cartography lab and we were a staff of seven faculty members, three lab assistants and less than 50 students enrolled for the postgraduate classes. In the apparent air of quiet, the sounds of discontent often simmered in the department. I chose to steer clear of tempers and camps. I was keen to create a niche of my own.

I opted for an office on the ground floor for two reasons: first, a ground location gave me the freedom of access at any time and day without needing the keys to operate the shutters on the first floor; second and equally important, the windows of the room opened to the lawns, with the *shisham* in close sight. My need to be with nature is addictive.

When I settled in my room at DSchool I felt that, I had achieved, what I had to in the material sense of being a success. I craved for little more than to pursue some ideas about geography.

I looked long outside the window and asked: Who was the first geographer in India? When did the study of geography start in India? How did the discipline evolve? Where are the departments of geography in India today? What are the preoccupations of geographers working in them? At what forums do they meet and interact? What is their specific contribution to India? What is the future of geography in India? A nagging thought: Where do I as a professional stand in this discipline? To my surprise, if not dismay, I found that answers to the above questions were fragmented, incomplete and far from satisfactory. This heightened my concern.

Concern for geography in India

To locate the discipline led me to search on how the discipline had evolved. India has a tradition of knowledge creation and none can deny the presence of illustrious universities like Nalanda, Takshila and Vikramshila. Geography must have ensured its presence in some way at these places. My regret is that I am handicapped in the reading or writing of Sanskrit, Pali or even Persian. This has constrained my researching the ancient or medieval texts. I have often wondered why geographers versatile in these languages, apart from the likes of Syed Muzaffar Ali, the renowned author of *Historical Geography of Puranas* (1966), cared little to throw light on the state of geography in historical times. This meant that I was left to depend on the colonial antecedents of the discipline.

Though the past was important, my clear interest was in the here and now of geography as a discipline in India. This took me to an interesting collection of facts and figures: addresses of the departments of geography, number of students taking degrees in geography, number of doctorates and their research interests, books and articles published by geographers, and activities of geographical associations in India. A host of such parameters were searched to understand the where and what of the status of geography at present.

I recall what the anthropologist C.J. Geertz had written: "one does not have to know everything to comprehend something." Yet because my data began to reveal a worrisome state of the discipline I wanted to verify my findings from different perspectives and in a variety of ways. I began to telescope on the role of geography in public policy and the inroads that technology was making into the subject. As if this was not enough, I took on the labour to cross compare geography with other social sciences, particularly economics and sociology in India.

I was barely putting all these facts and figures together when G. Krishan (GK), president of the National Association of Geographers, India, approached me with a request to collate the presidential addresses of the association for its forthcoming annual conference to be held at the University of Sagar, Madhya Pradesh in 2001. He had only one condition, that the book be released on that occasion. I had just six short months to race to the destination. GK is a man of timely work schedules. The elegant Seiko in my room at the DSchool was a gift from him. He is a man weighed with words and postures and could well have been an ambassador to another country. My writing of *Voice of Concern* had brought us close intellectually. The fact that he has written the foreword to all my books means that our mutual obligation remains steadfast.

I was going to make the most of GK's offer and set out a mandate larger than the one handed over to me. I decided to gather the presidential addresses, augment each of them with a brief biographical sketch of the president concerned, and chart the journey of the NAGI within the context of evolution and nature of geography in India. You see it was clear that the presidents and the association are intertwined in the scenario of the discipline. The trio: the geographers, their association and geography became the focus of my research. I stated what I discovered. Here are some excerpts from the *Voice of Concern*:

> ...the overall productivity reflects a weak intellectual interest and stamina of geographers in India. The normal requirement for a research publication in a journal in India has a word limit of 3,000 to 5,000. To write even this seems a daunting task for the majority. The per capita annual output of faculty at the departments of geography in Indian universities is less than 0.3 research articles. If the faculty in colleges were added to this, the figure would fall to a mere decimal.... The first doctorate in geography in India was awarded in 1940. Up until 2000, a doctorate had been awarded to 2,132 geographers in India. I have read the examiners' reports of over 40 dissertations. It seemed rather clear that more students were passed in a mood of compromise rather than conviction. The menace gets stronger when one detects that a collusive, though subtle, trade operates where supervisors are in league to mutually pass or get their own candidates passed. The situation does not bode well for the future.

I circulated my concern, mainly through a series of publications, as listed below:

Geography in India: A Future with a Difference (Book) 1998
Indian Geography: Tasks Ahead (Report) 2000
Creativity: Future of Geography in India. (Article) 2000
Voice of Concern: Geography in India (Book) 2002
Geography in India: A Languishing Social Science (Essay) 2004

The writings travelled fast even reaching non-resident Indian geographers. I received phone calls and letters, and these publications drew dozens of reviews in leading geographical journals. Today the publications are listed as readings in many colleges and university departments across India. I doubt if any book on geography in India has stirred the attention and interest of audiences as much as *Voice of Concern*. I can think of the following reasons for this response: First, scholars were feeling the slumbering state of geography in India and found their voice in this book. Second, the 22 presidents of the National Association of Geographers, India were among the known geographers in India and so took an avid interest in their book and thus participated in its circulation. Third, NAGI has a large pool of members which meant that the book had a captive market. Fourth other geographical associations in India took a close read of the book as a model to set out their own self inquiry

Languishing received a nationwide response and stirred the conscience of geographers not just in India but also in other countries. "In France, the situation is the same. Geography is not popular here, few students are enrolled, fewer books are published. On average students and researchers in geography are less bright than in some other disciplines." (Landy, 2005.)

Almost all the reviews were extremely laudatory, yet I cannot deny that while I was critiquing the state of geography in India, some geographers were criticizing, not critiquing, me. Among the papers submitted at a symposium on teaching and research in geography at the Centre for Earth Science Studies, Thiruvananthapuram (2012), I read one by S.K. Aggarwal, which stated, "… the anxiety regarding the state of geography in India has been expressed by many… and lately by Dr Anu Kapur in a widely discussed (rather criticized) paper in Economic and Political Weekly."

Ravi Singh, faculty at Banaras Hindu University, had the backbone to write this down rather forcefully: "That is how Kapur's (2004) views *were not liked by most of the professional Indian geographers* who feel over-satisfied with whatsoever they have done, without any inkling for improvement… No one can deny that she has invested [a] good deal of time and energy and brought out a formal perspective which was *so far have not been* [sic] analysed by any Indian in such vigorous manner…. This article is a successful one. It [has] created an atmosphere in departmental corridors, tea-tables and similar places to at least briefly discuss and change viewpoints around the problem where is geography in India today?"[1]

Interestingly those who criticized me way back in 2002 are the same scholars who now in 2020 go around drumming that something needs to be done urgently about the worrisome health of the discipline. They argue for an urgent need to eradicate the maladies of the discipline in terms of modernizing the contents of syllabi, improving the quality of teaching and inculcating values within the profession. I would not be exaggerating if I said that a serious discourse on the health of geography in India assumed a regular feature among academia after the publication of *Voice of Concern: Geography in India* and *Geography in India: A Languishing Social Science*.

The titles of my writings reveal that there was a subconscious design at work when I was charting the status of geography in India. I begin with my first publication titled *Geography in India: A Future with a Difference*, followed by *Voice of Concern*, and I put a kind of finish to my publications on this theme with the essay *Creativity: Future for Geography in India*.

I became increasingly convinced of the need for creative thinking in geography and felt that this alone could lead to new knowledge or new ways to look at old knowledge. I was so driven that I published a diagrammatic model of spatial creativity which, in a way, summed up my definition of geography. I was greatly influenced by the Maltese physician, Edward De Bono's work on lateral thinking and so was convinced that the plasticity of the mind can be reconfigured. I strongly disagree that creativity belongs exclusively to the specially gifted, a minority sometimes called genius. In a seminar at the Centre for Regional Development, Jawaharlal Nehru University, Aijazuddin Ahmad and I locked horns on this issue. He observed that creativity was an inherent endowment born with trait, while I fiercely stressed what the illustrious German philosopher Friedrich Nietzsche wrote:

> Do not talk about giftedness, inborn talents! One can name great men of all kinds who were... little gifted. They acquired greatness, became 'geniuses' (as we put it), through qualities the lack of which no one who knew what they were would boast of: they all possessed that serious-ness of the efficient workman which first learns to construct the parts properly before it ventures to fashion a great whole; they allowed them-selves time for it, because they took more pleasure in making the little, secondary things well than in the effect of a dazzling whole.
>
> Friedrich Nietzsche, Human, All Too Human:
> A Book for Free Spirits

The past is over, gone and cannot be revisited. The present is the only space where we can gain insight for any foresight. Eager to project into the future I coined the word geoimagining to emphasize the need for geographers to develop their power of creative imagination to foresee the problems likely to confront the world and figure out possible geographical solutions. The essay "Geography in India: A Languishing Social Science" concludes on this note:

> ...if geographers want to amount to anything useful, they cannot help but respond directly to the many critical real life problems which con-front India. The India of [the] 2000s will be markedly different from that of the 1950s. Then what would the India of 2050 be like. Geo-imagining the future, geographers need to align their research to the emerging demands of society. There is an urgent need to reinvent our commitment.

Strange that while I chose to critique the discipline, I never could get around to critiquing the geographers. I seem to have made two separate baskets; one for geography and the other for geographers. It is not that I did not know of the geographers; nor was I in fear of them in any way. Why then did I hesitate to critique them? I just felt that it is important not to think about people in a hurried way – I did not want to mistake them, so I chose to be cautious and saw no reason to be critical either, for each of us is trying to do the best we can at our level of consciousness. The line between light and dark remains grey in life.

As for me, the experience of writing on geography and geographers in India was indelible and inescapable. This research allowed me to introduce a course on geography in India at the Department of Geography, DSchool when the syllabus was revised in 2018. Ours is the first department in the country to offer an independent course on the discipline in India. I feel that by its inclusion I have managed to sustain continuity about the deliberation of the status of geography in India.

The research on *Geography in India* not only brought me into contact with the thoughts and mannerisms of many geographers, but certainly it seems that it was I who heard the core message of the *Voice* loud and clear. I did not want to spend time battling what ails geography; I essentially wanted to *do* geography. It is for this reason that after spending a decade on the study of geography in India I set to work on *Geography of India*.

On geography of India

Geography inheres a *dhamma* – in the sense that those who have a deep interest in the earth must have a special concern for its wellbeing. I had been profoundly impressed by George Perkins Marsh's book, *Man and Nature of Physical Geography as Modified by Human Action* (1864). Marsh was concerned with the environment not simply as an object of study but one that creates pressing practical and ethical issues for humanity. Written equally stirringly was Rachel Carson's *Silent Spring* (1962), which provided a telling description of nature's backlash to chemical pesticides. I cannot, likewise, forget *Small Is Beautiful: Economics as if People Mattered* (1973) by Ernst F. Schumacher, which made a case for nature friendly solutions to human problems. I also read *The Limits to Growth* (1972), a report by a group of economists which highlighted the criticality of exponential economic and population growth in a context of a finite supply of resources. The landmark report of the United Nations Conference on the Human Environment, published in 1972, was of no less interest to me. By the 1990s when I began to think of an appropriate research problem for my doctoral thesis, the world was in the midst of an environmental crisis. Under the influence of such a frame of ideas I was most inspired to research the human impact on environment.

Kashmir Valley: a paradise in peril

Kashmir is one of the largest and most populous valleys in the Himalayas, and one known for its enchanting natural beauty: *Kashmir – bi-nuzir* (without an equal), *Kashmir – junat-puzir* (equal to paradise).

Today the region is entangled within a serious political turmoil, but *Paradise in Peril* took a different tangent and argues that while political situations can be resolved it would be hard to stem an ecological ruin. My thesis pronounced that human greed and associated actions take a paradise to a state of environmental peril. I structured my argument into three successive sections titled – Genesis to Decline, Ecological Degradation and Perilous Cycle of the Paradise.

Adopting an evolutionary approach, the first section Genesis to Decline, journeys the sequential occupancy of the valley across ancient, medieval, colonial to post-independence periods and nails the initiation of ecological ruin to the colonial period. The British set out a devious blueprint of exploiting natural resources to realize their design of imperial expansion and control. In the post-independence years, in the name of national development, the exploitation of nature continued unabated. The resultant consequences of this destruction are described in the second section of the book titled *Ecological Degradation*.

An environmental system is in peril when the equilibrium of nature's arrangements are jeopardized. The peril in this case was presaged by a host of warnings: the shrunk Dal and Wular lakes, the polluted Jhelum, the stubborn pests in apple orchards, the gregarious spread of weeds, the floods above danger mark when rainfall is below normal and the rock piles as landslides blocked all major roads. The series of links – de-vegetation, soil erosion, siltation, choking water bodies, flood and drought – are underway in the valley. Pollution, depletion and degradation – these are not separate ills. The finale of all this is elaborated in the third section of the book titled the Perilous Cycle. The telling message is that when the environment is tampered with and its thresholds of resilience are crossed, then a series of backlashes are unleashed which create the syndrome of an ecological peril.

The book provided a graphic description and in-depth analysis of the ecological crisis and forewarned about the impending climate change and the scope for unprecedented large floods engulfing the valley. The latter seemed such a clear possibility, rather a certainty, and I became so worried that straight after my book I sat to write a research paper titled *From a Lake to a Lake: An Environmental Prognosis of the Kashmir Valley*.

In this research paper I built a scenario of a sequence of three lakes in the Kashmir Valley. I called the first a geological lake, because the Kashmir Valley's elliptical area of 13,000 km² was once covered by the sea of the Tethys. The valley geologically thus has an aquatic origin.

I addressed the second lake as a geomorphic lake which was created when glaciation and de-glaciation filled the valley with sediments so deep and thick that the Jhelum River was unable to carry its load out of the narrow gorge it encountered at Baramulla. The back flow of Jhelum River resulted in the creation of a huge lake. This process happened 2.5 million years ago during the Pleistocene epoch. The fluvio-glacial deposits locally known as *karewas* are evidence to this process within the valley.

Both the geological and the geomorphic lake are results of a natural process when human agency was nowhere on the scene. The worrisome one for me is the possibility of a third lake which, given the alarming ecological deterioration, would develop by the end of the 21st century.

I address the third as an anthropogenic lake. Unlike the earlier two, the third lake would be a backlash to human action of deforestation, followed by siltation flooding and landslides along with other ecological processes like pollution and climate change. The lake would impact the lives of millions of inhabitants of the valley. It would drown its most fertile and prized agricultural lands and submerge several towns and villages. In all probability, the lake would suddenly form without notice, and human beings would be forced to abandon their homes and move out. After armouring this forecast with rigorous data and analysis, I concluded that "Kashmir Valley is threatened with the possibility of being drowned in a human created lake by the turn of the 21st century."

My book and the research paper were published in the mid-1990s. The premonition they carried manifested in 2014–2015 when headlines in many local and national newspapers carried the following: Kashmir Valley turns into a lake; several thousand villages across the Kashmir Valley had been hit by a devastating flood, in many places of Srinagar's neighbourhood, the water was about 12 feet deep, submerging entire houses. Echoing a similar situation, the headlines that reappeared in 2018 were : "Kashmir floods: panic spreads in the valley as water level rises" "Kashmir Valley drowns" "Valley submerged."

These are just a few examples gleaned from many media reports in the *Kashmir Times* and *Times of India*. It seems that the events stole some pages from my book. My thesis was endorsed. We surely could have saved much if the peril had been heeded.

I am tempted to reproduce here a recent article by Dinesh Singh (*Daily Mail* 16 March 2020).

> … Some years ago, Delhi School of Economics' Professor Anu Kapur had written a paper on the vulnerability of Srinagar to flooding. Her arguments were very rational and based on much data. A few years after her paper was published, she may well have acquired a Nostradamus-like persona, since the disaster, as predicted by her, did occur. The unfortunate part is that her work was never brought into discussion before or after the actual occurrence of the disaster.

Yet all is not bleak.

Won: an ecological trial in the High Court of Delhi

In the hub of Central Delhi's densely packed commercial and residential area of Karol Bagh lies a small 24-acre park, called Ajmal Khan. My home is located opposite the park. The deteriorating ecology of this park thus became my field of study. Since the area of the park is small, one could virtually traverse, photograph and map each inch of the ground. The symptoms of the ecological decline be it the poorly maintained lawns, the stunted shrubs, the lopped trees or the squalor and filth were all too conspicuous. Most disturbing were the slithery concrete structures that were spreading their tentacles to overtake the park.

This research, under the title *Ecological Perspectives of City Greens: A Study of Ajmal Khan Park, Delhi*, was published in 2002 by the Human Settlement Management Institute, a wing of the Housing and Urban Development Corporation (HUDCO), government of India. As in almost all of my works, I seem to have a proclivity to envisage the future. In this study too, I wrote a chapter titled, "Ecological Prognosis." Herein I forewarn, "if tangible actions are not taken soon, commercial interests will swamp the park with concrete structures." This threat was about to materialize.

A private consortium called International Food Plaza, in connivance with the officials of the Municipal Corporation of New Delhi – who, ironically, are the officially designated caretakers of the city parks – started constructing a banquet hall within the heart of the Ajmal Khan Park in 2004. As soon as I came to know of this ill-advised encroachment, I forwarded my HUDCO report to Paryavaran Avam Januthan Mission, a non-government organization devoted to the cause of environmental issues in Delhi.

We filed a Public Interest Litigation Case in the High Court of New Delhi in 2005. It was a legal battle over many dates of arguments in the High Court. The final judgement pronounced on November 2009 ordered that, *concrete structures and all other illegal activities that were spoiling the greens to be removed with immediate effect.* Under court order the banquet hall was demolished and green areas restored. (High Court of Delhi, New Delhi, WP(C) No 6950/2009 and 2309, 3925, 4894, 5329/2009). The observations from my HUDCO report formed a crucial part of the Delhi High Court judgement.

In the preface to the HUDCO study I emphasize that

> From indifference to action to violence, there are many ways to deal with the deteriorating ecology…Somewhere in the middle lies research and contemplation – if concerns raised can persuade some people in authority to take action, much could be accomplished; decisions for action, after all, draw their strength from the nourishment of thought. Action requires effort combined with hope. Pessimists I know would cringe at this hope, the cynics would laugh, but I am realist enough to know that hope sustains life …

Each life is precious; therefore how lives could be saved demands a serious research inquiry.

Disasters in India

Just as I was rejoicing at the favourable judgement from the High Court, which saved the Ajmal Khan Park, I began to heed another loud and this time nationwide alert: Andhra Pradesh cyclone 1990, Uttar Pradesh floods 1993, Killari earthquake 1993, Gujarat cyclone 1998, Orissa cyclone 1999, drought 2000, Bhuj earthquake 2001… The media and television reports were filled almost on a daily basis with the number of killed and affected and the scale of damage and destruction.

I was keen to take a count of the toll and damage and came to learn that in the period 1982–2001 as many as 59 million people were affected by disasters in India. This approximates the total population of the UK. The average number of annual deaths in disasters during these years, was four times more than the average for Asia, and eight times more than the world average. India's top ranking in the number of people killed and affected by disasters worldwide did not go unnoticed by me. The near endless cry of disaster: the relay of tragedy, sorrow and despair beckoned research.

Let us recall that at the end of the 1990s and early 2000s India was marked by an indifference to disasters. I brought forth this apathy in my essay "Insensitive India: Attitudes Towards Disaster Prevention and Management" *Economic and Political Weekly* (2005). In this essay I documented insights into the lax scenario of focus on disasters at a national level. In 2003 India had 43 ministries but none which dealt exclusively with disasters. It was as late as 2005 that India set up its National Disaster Management Authority. It was only in this very year that its first Disaster Management Act was legislated; all this delay despite the fact that India had been a signatory of the International Decade for Natural Disaster Reduction, beginning on 1 January 1990.

In the writing of this article it was more than evident that the country had ignored and was grossly delayed in its obligations. My essay concluded that

> Before we again start crying hoarsely that millions have become homeless and infrastructure worth billions has been destroyed we need to ask where in the first place the houses, the hospitals or the services are? Where are those politicians, administrators, planners, media and academicians who promise and pledge to care for the nation? By disregarding that each life killed and each home destroyed could have been saved, this insensitive cohort of India has made India vulnerable. They are the disaster!

I did not want to belong to the insensitive cohort of India. I was convinced that the situation of disasters could be changed only if one has a clear comprehension of the problem to be dealt; shooting darts in mid-air and blaming nature as reason for disasters could get us nowhere and would bear only bitter fruits. My empirical and theoretical contribution to knowledge became fixed on the study of disasters for almost a decade.

What are the kinds of disasters that India faces? Why does India figure in the top ranks in world disaster inventories? Why do so many disasters recur in India? Which regions are hotspots of disasters? Who are the people killed and affected? What is the response of the people and authorities to disasters? How best can India be made disaster free? The apparent simplicity of these questions set me to the task of seeking an answer. I built on these critical issues consistently and persistently through teaching and research. The course on Man and Environment that I taught to my postgraduate class carried a caveat which covered issues pertaining to disasters. In my teaching I consciously focused and elaborated on the parameters of this module. The result was that a batch of dedicated and enthusiastic students were attracted to pursue their MPhil and PhD on disaster related topics. Here is a selective list of topics covered by student research:

Vulnerability and Response to Disasters in the Andaman and Nicobar Islands
Fire Disasters in Delhi: A Spatial Analysis
Bhuj Earthquake: A Geographical Study of Disaster and Response

Impact and Response to Bhuj Earthquake in Rural and Urban Areas, Kachchh

Drought in Baran District of Rajasthan: A Study of Vulnerability and Response

Vulnerability to Flood Hazards in Majuli, Assam

Response to Drought in the Rain shadow Zone of Maharashtra

Disasters in the Slums of Kolkata

Disasters in Greater Mumbai: A Geographical Inquiry

Response to Hazardous Industries of Delhi: A Geographical Inquiry

Disaster and Response: Killari Earthquake in Latur District, Maharashtra

Disasters in Bhubaneshwar City: A Geographical Interpretation

Disasters in Indian Railways

Some of the dissertations were outstanding in their quality and gained immediate publication. The one on the sequence of disasters in Andaman and Nicobar Islands was picked out rightly by Routledge. This was a doctoral thesis. It instilled a sense of pride.

Grim reality

Since all these research studies were conducted under my supervision, they carried a kind of similarity with each other in substance, spirit and style. Five students in the list above were eager to spread their research beyond the confines of a dissertation in a department cupboard. These five students and I decided to put together a book which we titled – *Disasters in India: Studies of Grim Reality*. By the nature of its origin, the book is in the form of in-depth case studies. The various chapters adhered to the call of the title of the book: "Delhi: A disaster prone national capital"; "Greater Mumbai: Teeming with disasters"; "Kolkata: A city on the brink"; "Bhuj Earthquake: Rural–urban differentiation"; "Problematic City Space: Relocation of hazardous industries"; and "Indian Railways: On wheels of death."

Since it turned out to be the combined work of a supervisor and her students it was only fair that all were named authors of the book. Each of us felt pride in signing the contract of consent with the publisher and receiving the royalty cheque as and when the publisher obliged. It felt good.

The student theses and dissertations literally became my eyes and ears as to how different types of disasters were manifesting themselves in various parts of India. Cities turned out to be more disaster prone than other areas; the largest number of victims were the poor people of India; and among the poor, women and children suffered the most. I could see a clear disaster divide. Research provides reliable data and concrete solutions instead of opinions and hypothetical conjectures. After all, disasters cannot be managed on guess work or unfounded conjecture. Policies, programmes and governance rest on research.

While I was reading, learning, and teaching the course on disasters my students frequently lamented that research on disasters on India is difficult to find; it is scattered and fragmented. Let us not forget that we are talking of the situation nearing the end of 1990s. At that time as a subject, disasters were not taught in schools or colleges, there was no National Disaster Management Authority to speak of, nor a Disaster Management Act.

On disaster research

Research on disaster research thus became an urgent call. India just could not afford to wait for millions more to die before acting. Quick access to the available research emerged as almost a necessity for the country. Taking a cue and with the support of my students I tied together all the research on disasters in India from 1830 to 2006. In line with its contents and contexts, I titled the book *On Disasters in India*. I cast a wide net collecting and covering all possible disciplines and outlets of serious research. An enormous quantity of material was unearthed, shovelled and sifted before a reference could be mined. The first tracks are the hardest to build. I hunted bibliographies, sorted bookshelves, tracked unpublished reports, listed doctorates, sifted journals and noted the references in others' references. My appetite was insatiable.

The cover of *India Today* is one case. This magazine flagged its first edition in 1978. An estimated four million people subscribe to its English edition. I wanted to identify how many cover pages of the magazine had focused on disasters in its 25 years of publication from 1978 to 2002. Disasters had been a theme of concern on only 22 cover pages! This was the result after sorting a stack of 750 issues of this magazine.

The degree of my obsession is apparent in another search. To flesh out 113 doctorates on the theme of disasters in India, called for a careful reading of the titles of 1.8 lakh doctorates in the Bibliography of Doctoral Dissertations, published by the Association of Indian Universities! It was eye-straining but exciting.

Wherever I looked I saw a paucity of work published on disasters and this inspired me to pursue wide and deep. When I had netted 4,004 captions of research on disasters I knew I had to put draw a line. The number 4,004 was not the stoppable high, but what I found was that the time spent chasing one more research piece was not yielding any results. This was not the end. This pile of collection of research needed also to be researched. After all my role was not that of a bibliography maker but a researcher on disasters.

Here is where the Excel program lent itself as an excellent tool. Once the 4,004 pieces of research were keyed into an Excel spreadsheet, I had before me the makings of big data and here is how: each research piece carried a year, type of publication, details of the author and so much more that could be picked alongside. The related queries were many: Who were the pioneers of disaster studies in India? Which among the physical and social sciences harbours a greater interest in research on disasters? Has research on disasters in India taken scholars far and wide or have they confined themselves to a few spaces and places? What are the factors that have led to the waxing and waning of interest in the study of disasters in India?

Excel lent the ease and efficiency to rank, organize, classify and sieve the big data in such a variety of ways that I was able to graph the researchscape of disasters. All the research entries were examined in terms of their basic nature or type, occurrence, cause, impact, response, forecasting and management. This analysis revealed that research on disasters falls into three historical periods: Colonial (1830–1947): a Phase of Awareness; Post-Independence (1947–1989): a Phase of Indifference; and Globalization (1990 onwards): a Phase of Recognition.

My two books, *Disasters in India: Studies in Grim Reality* and that titled *On Disasters in India*, became a valuable springboard to hone my more pressing questions about disasters in India. The dominating thought that nagged me frequently carried a number of questions: Are disasters natural? Is physical geography the reason for disasters in India? Are cyclones, earthquakes and floods really responsible for the killings and damage? Is nature the cause of disasters or are we running a cycle of blame game? What had to follow was my book *Vulnerable India: A Geographical Study of Disasters*.

Vulnerable India

Vulnerable India came to me as the grand finale of a symphony of sorts. It seemed I had carried an obsession with the disasters theme. My lectures, case studies, perusal of literature and the research I had supervised on this theme soaked me with data, thoughts and ideas on disasters. I knew that I wanted to write about disasters, I also knew with great clarity that I wanted to *de-myth* the idea of natural disasters. I was greatly impressed by the Scottish philosopher David Hume, who observed that it is unreasonable to conclude that merely because one event precedes another that thereafter one is the cause and the other the effect. This conjunction may be arbitrary and casual. There may be no reason to infer the existence of one from the emergence of the other. I felt that this is exactly what we at large have been doing in the case of disasters. Just because the physical event preceded the disaster we were drawing a serious category mistake by dubbing the reason for disasters as natural. Yet I needed to set my argument and prove my thesis to be true.

With this *junoon* (obsession) came a spontaneous willingness to lay everything else aside, to retreat from all manner of immediate interactions from the world and it is here that a timely fellowship at the Indian Institute of Advanced Studies, Shimla proved to be a boon. The charter of the institute states "…it was to remain a residential centre for free and creative enquiry into fundamental problems of life and thought." *Vulnerable India* seemed to have found a perfect ambit of research. I took on no outside assignment or pressures when I sat down to write *Vulnerable India*. I also allowed no one to influence my context and content. I structured my argument into three sections of Fact, Response and Reality.

Fact

To prove that it is a fact that India is a land teeming with disasters and to capture its spatial symptoms I coined the word disasterscape. It was my study of the German concept of *landschaft* or landscape that led to the possibility of me being able to add a new word, disasterscape, to the dictionary.

I wasted little time and moved to my urgent concern: Why does a disasterscape so often scar the face of India and how do I break the false connect between natural and disaster? It was clear that the natural are those processes whose origin, evolution and functioning are independent of humans. Encompassing the realms of land, water and air, these are all generated by and tied to physical processes. Because of this origin I decided to replace the word natural with the word geophysicals. My idea was not the casual adoption of a synonym. The intention was to ensure clarity by underlining that geophysical are an integral part of the functioning of the earth's system. This being the case nothing can stop them. They are what they are, and will always be so! Take the case of earthquakes; they will arrive when the tectonic plates of the earth shift and there is nothing human beings can do to stop their arrival!

Since I did not see the role of geophysical in causing the disasters I prefixed the word ascribed to them. This was intended to convey that disasters are incorrectly attributed to geophysical, which is merely a misplaced ascription. It was important that this clarity was communicated by the choice of appropriate words. I then proceeded to analyse the temporal-spatial character, annual and seasonal behaviour and frequency of occurrence of the ascribed geophysicals. This was because even if they are just an ascription it cannot be denied that the disaster occurred. For this I selected the following 16 geophysicals: (i) earthquake, (ii) cyclone, (iii) heavy rain, (iv) drought, (v) cloudburst, (vi) thunderstorm, (vii) gale, (viii) squall, (ix) flash flood, (x) flood, (xi) hailstorm, (xii) snowfall, (xiii) cold wave, (xiv) heatwave, (xv) drought and (xvi) dust storm. Since my focus was to prove that disasters are not natural, for obvious reasons railways, fire or industrial and other such disasters were not to be taken into consideration.

Considering the probability that some among the 16 geophysicals may not occur in a given year, I collected data from 1977 to 2002. The assessment was that packing together information for 26 years would display a firm temporal-spatial pattern of disasters in India. Since earthquakes and cyclones are distinguished by high-magnitude but low frequency of occurrence, in their case a 100-year period was considered desirable. The data netted 6,017 disasters ascribed to the 16 geophysicals over the years. Clearly I was dealing with what can be defined as big data. Since my lens was geographical, this meant that the map became an important tool to convey the message of my research.

India is the 7th largest country in the world with a total area of 3,287,263 km². The choice of scale for making a map is critical. Selecting a small scale would depict a rather generalized state of disasters, while on a very large scale the grain size would have become so fine that the overall picture would turn hazy. I could have worked through India's 28 states and seven union territories, or the 5,453 tehsils or approximately 6,000 blocks. But I found these ends of the spectrum unsuitable. In a bid to view the forest and yet not lose the trees, I rooted my inquiry at the district level.

Logically it seemed the optimal choice to use the district as the basic unit of administration; all the major departments of state government are represented at its headquarters, central government channels all its programmes and schemes through the district administration, and a district is placed under the care of a member of the Indian Administrative Service attached to a specific state. By and large, a district is marked by a fair degree of homogeneity in resources and development.

A daunting roadblock was that the data for the 16 geophysicals did not come ready-made by districts. The location of epicentres of 542 earthquakes was listed according to latitude and longitude and not by the name of the district. Similarly the 236 landfall sites of cyclones were mentioned either by the name of villages or location in proximity to an important town. The spatial spread of geophysicals such as snow, or heatwave or flood was reported often by area or region and not by districts. To build a directory of addresses for the disasters during the 26 years under consideration created a challenge to decode as accurately as possible the locational dimensions of disasters. To scrutinize, sort, separate, key and map all such disasters at the district level for India was indeed an eye-straining and back-breaking chore. I had a few passionate students who were admirably helpful.

There were other challenges. The 26 years selected for inquiry spanned across four inter-decadal intercepts. The administrative map of India carried 356 districts in 1971, 412 in 1981, 466 in 1991 and 593 in 2001. To avoid duplication, confusion or misidentification, the districts of 2001 were adopted as the base for data processing, mapping and analysis.

I realize that every solution solves a problem yet creates its own set of problems. The fixing of 2001 as the year to work involved dealing with three prior censuses on issues like the division of one district into a large number of districts; the merger of parts of two or more districts; and other such territorial readjustments.

While the spatial mapping of disasters was organized within the framework of all the 593 district boundaries as in Census of India, 2001, yet I found that I was unable to discuss the patterns using such a large list of district names. Therefore, for a meaningful presentation of my findings, I had to find a meaningful regionalization scheme. The Census of India publishes its physiographic divisions of India with boundaries of a district as the basic building block. This to me offered a manageable nested hierarchy of four macro, 28 meso and 101 micro-regions for India. These were definitely a more manageable number than the 593 districts. As an additional virtue, the names of regions carry a closer affinity to the place and its description. It is through these regions that I built my scenario of disasters.

After an analysis of each of the 16 geophysicals, be it cyclone, flood, drought, earthquake or others, a question that lurked was how they all congregate and fold one upon the other, and thereby what does India's holistic picture of disasters that are ascribed to geophysicals look like?

Packing all these geophysicals together was neither easy nor simple. While a query on the geographical information systems (GIS) software could simply layer one on top of another, a decisive dilemma was how to equate a lightning with a cyclone, an earthquake with a flood or, for that matter, a heatwave with flash flood. Each is so different in their origin, spread and impact.

A solution was sought by assigning a separate weight to each of the 16 geophysicals. In the present case, the weight of a geophysical occurrence was calculated by taking into account the ratio of those killed per disastrous occurrence. Therefore, in the case of cloudburst, its 23 occurrences with 192 killings gave the ratio of 1:8 or a weight of 8; in cyclone, the 91 occurrences and 43,389 killings generated a ratio of 1:477 and hence a weight of 477. Mapping this cumulative load for each district helped to discern the shades of disasterscape as it varied from one part of India to another.

Response

Having established the facts, the next logical step was to try to find out how, when, and why the idea of natural disaster originated and became entrenched as the dominant response to disasters. I traced and referred the reason for this paradigm of the naturals to three successive perspectives which took shape during the traditional, colonial and post-independence phases.

While the section on Fact led to my involvement with numbers, statistics and GIS, the section titled "Response" burrowed me into tracking popular narratives and reading philosophical texts and government reports on India. A wide read across sociology, economics, philosophy, religion and history was essential. This is where being at the IIAS was most providential. The library here is rich in books and research journals of the social sciences and humanities. Fellows could borrow 50 books at a time! Had I not indulged and reflected on this literature I would have wrongly concluded that while in all three phases nature was held responsible for disasters, the reason for this attribution differed in each phase. My perusal of relevant literature enabled me to sense that within traditional parlance, disasters are viewed as the wrath of the divine, during the colonial days these were attributed to vagaries of nature, and during the post-independence days it was the apathy of the government which found it convenient to propagate this thesis.

I could clearly recognize that the paradigm of blaming nature was inherent in the 18th century science of the Europeans and this was couriered into India by the British. But I dismiss the fact of holding the British as responsible for the dismal state of affairs by reiterating that the West did find ways to deal with disasters, whereas in India the entire scheme of organizing, forecasting, and distributing funds for relief during the post-independence phase continues to ascribe the agents of death and destruction to geophysical phenomena. In fact, I prove that in the over 70 years since independence, India's response to disasters remains locked into a blame game of nature with total indifference to the plight of its people, and denial of political and administrative responsibility. The objective of my research was not to point fingers. I was keen to learn who or what is responsible for disasters in India if not the naturals? Why are its people subject to a catena of disasters? Why does India figure time and again among the top ranking countries for disasters?

This moved me to the third section of my book which I boldly titled: Reality.

Reality

In this section I used every weapon to dislodge and undo the myth of the natural and proposed the argument that it is a Vulnerable India that is the reason for disasters. Who are the vulnerable? How does one define the vulnerable? What is the spatial pattern of their distribution? What is the size and composition of the vulnerable population? How can they be measured and mapped? How do they define the contours of India's disasterscape?

India is a billion plus country. It harbours extremes of density, diversity and differences. This means that while everything may be true for a part of India, it may not be true for all of India. Hence at the outset I needed a mutually comparable scale of analysis for both disasters and vulnerability and it made sense to keep the district as the common choice for the spatial unit of analysis.

The core issue was how to measure the vulnerable? The concept of vulnerability is abstract and elusive. I would need strong surrogates as indicators to build an index of vulnerability for disasters. I was halfway through this exercise when I realized that I had made a serious blunder while selecting the indicators of vulnerability.

Here is where I faulted: I had selected 14 indicators to measure vulnerability but among these 14, I took one which should never have been there. This was the data on the composite loading of the geophysicals. You see, my premise was that naturals are not the reason for the vulnerable, yet it was I who had placed naturals as one of the indicators of vulnerability. I had made a serious category mistake of tying in cause and effect. It took me many months to undo my own intellectual opacity.

In the final I needed indicators that represent and synthesize the properties of vulnerability. I wanted the indicators to net in a meaningful range of dimensions of vulnerability. It was also important that the indicators drew their strength by belonging to authentic sources. Further, I wanted to ensure consistency in respect of their information and that all data was available at the scale of the district.

It was a tall but not impossible order and finally, after many searches, screeches and screams I was able to sift 13 indicators to measure vulnerability. These were then grouped into three components. For the component of disadvantaged people I selected five indicators, namely, illiterate females, marginal workers, agricultural labourers, scheduled caste and scheduled tribe population, and the disabled. The criteria of fragile living was measured by four indicators: households below the poverty line, households living in *kutcha or semi pucca* houses, households lacking specific assets and the infant mortality rate. To measure lack in services the three indicators selected were un-irrigated crop area, rural–urban distancing and number of persons per doctor. The idea was to measure the resilience and capacity of people to protect themselves in the wake of disasters. It was the synthesis of these three criteria with which I built the overall vulnerability index. I was satisfied with my choice as each of the three components captures a different dimension of vulnerability which is just what I sought.

I then decided to do the one thing that I had waited for all along: compare and contrast the map of the disasterscape vis-à-vis the map of vulnerability of India. What I expected was that the two maps would share a close overlap in the sense that the higher the vulnerability the larger the score on disasters. But what I found instead was this: wherever there is a disasterscape there is vulnerability, but vulnerability is more widespread than the disasterscape. Now this left me most bewildered.

When things do not appear right I have a strong tendency to tell myself "I must have done something wrong!" In this state of self-blame, I set out to recalculate the data and redo all the maps and there were miles of Excel sheets and hundreds of maps. I did this because I thought I had made a mistake, but since I didn't know at which stage I had erred I set out to recheck all the calculations and the GIS queries that I had generated. However, this did not lead anywhere. After all, doing the same thing over and over again does not deliver different results.

I was deeply anguished and disturbed yet did not balk. I set out to create clusters of vulnerability. By using the statistical package for social sciences (SPSS), a decision was taken that six clusters would be manageable for capturing the spatial reality of India. While each of the six clusters represented different shades of vulnerability what was more revealing was that only one out of the six clusters stood out as a case of low vulnerability. The description that India carried an island of low vulnerability amidst an ocean of vulnerability may sound a cliché, as was attributed to the spatial pattern of development in 1950s, but could well apply also to the prevailing scenario.

What was converted into my next dilemma was that when I pitched this low vulnerability cluster vis-à-vis disasterscape even this relatively low cluster was not free from disasters.

I was getting a bit sceptical and began to wonder if naturals are the dominant cause for disasters? Was I chasing a wild thought and being stubborn to my thesis? Yet I held steadfast. I asked myself how many people were vulnerable to ascribed geophysicals or so called natural disasters in India?

When the vulnerable population of each district was added, it totalled 511,028,530, making 49.68 per cent of India's total population. This is nearly double the size of India's population below the poverty line. What was revealed is that India's vulnerable people are in every nook and corner. Hence the map of disasterscape and that of vulnerability did not emerge in what had been my expected congruence about each other.

Three reasons came to mind to explain such a situation.

First, even low vulnerability is not low enough; it is only a relative category. Second, the areas of low vulnerability are not without vulnerable people. Finally, geophysicals are bound to manifest themselves in all areas, even in areas of low vulnerability, and not everyone is resilient to disasters.

Gripped by the state of affairs, I went ahead to emphasize that one should distinguish between vulnerability *simpliciter* and rendered vulnerability. Vulnerability *simpliciter,* like mortality, is an inescapable fact of human life. But to be killed due to rendered vulnerability is a different matter. It represents failure of the system to protect human life. It is with this reasoning that I insist on the need for a vocabulary change and argue that those who lose their lives in a disaster should not be referred to as those who died but be described as those killed.

Likewise I wanted to draw attention to the distinction between a cause and reason. Causes are discovered and reasons are investigated. Conceptually, reason implies agency and entails responsibility. Killing belongs to the category of reason, pointing to some responsible agency. When a house collapses and the inhabitants die, it has to be said that they were killed. A host of factors had allowed a house to be so fragile that it could not withstand the effect of a downpour or a cloudburst or an earthquake. Since this is a sure case of human irresponsibility and indifference, therefore the befitting phrase is rendered vulnerability. The word rendering means the making of and in this case it is the making of disaster. The phrase, Vulnerable India is not an uncommitted neutral description. It is a reasoned judgement of concern. Human at its very core, it awaits a humane response.

The writing of this book was an overwhelming experience. For me *Vulnerable India* remains the most important work I have done. It was not an easy write. I rewrote the draft several times and I confess that the version I submitted to the Institute of Advanced Studies, Shimla and that published by Sage bear only a nominal resemblance.

The book took a while to become known but soon came reviews arriving from different directions. The book is also listed as a compulsory read in the disaster management syllabi of the University Grants Commission; Panjab University, Patiala; Anna University, Chennai; Panjab University,

Chandigarh; University of Delhi; and several other institutions across India and is also required reading at the MIT. I was invited for lectures and discussions on the theme. I was keen to seed a disaster centre at the University of Delhi and had taken my blueprint to Vice Chancellor Dinesh Singh (2010–2015) but before we could embark on a serious discussion, a political change at the helm of affairs put the idea on a back burner.

I soon realized that while my lectures received applause and the reviews too were acclamatory, no one was really heeding my core message of disentangling natural from disaster.

Journalists and the media continue to address disasters using adjectives such as fury, violence and devastation caused by nature, army personnel continue to carry out mock drills as a way to prepare civilians for an attack by nature, while politicians announce relief and funds for the wrath of nature. Nothing has changed; what is more, in the teaching and research even my own fraternity continues to trumpet the hypothesis of natural disasters.

Perhaps what there is to see is given by what you see it as. I do not know if anyone noticed but I wanted to initiate a new line of inquiry and so ventured to change vocabulary. I wanted governments and policymakers to look at people afflicted as killed and not simply dead, disasterscapes being rendered and not just a chance happening. But no one in India seems to be heeding. While I am sure its time will come, After *Vulnerable India* I was kind of exhausted. Moments of sadness, apathy or despair would saunter in without warning. I repeatedly thought about vulnerable people and the possibility of disasters striking and wiping out so many. My father's cardiac surgery in 2006 had exposed me to the real vulnerability in life and how hard and vigilant one had to work to rebuild resilience. It took me over a year to put my dad back into the saddle of his work.

The three books *Grim Reality*, *On Disasters*, and *Vulnerable India*, are a trio that appear to be a high point in my career – in fact, it turned out to be the lowest phase of my life. I suspect that with the study of disasters, the theme of India's glaring disparity, and the presence of pervasive insensitivity had got to me. Though I continued to research sporadically on the old, orphaned and disabled – those I describe as the most vulnerable among the vulnerable – I was in a pensive mood and keen to search for less heart-wrenching themes. I needed to shake the damp settling on my shoulders. For an uneasy and uncertain period I continued to drift; I tried to map the terrain of Delhi using its high-rise buildings as measures of 3D mapping, dabbled with issues of sustainability, wrote a few book reviews and produced nothing.

I longed to bask in more peaceful terrains so I took a complete U-turn. I opted to enjoy and capture the beauty, diversity and distinctive capacity of India. It is here that the heart of geography, pure and simple, beckoned me. In the next two books I pay homage to the geographical knowledge of place. All human life is place specific. Nothing mattered more than coming back to place.

On place

From nursery to the Senior Cambridge Board Exam I went to only one school and studied at only one university up to doctorate from graduation. I moved from assistant professor to a full professor working at one university; I have lived in one ancestral house from my birth to the present. I obviously stay put in one place. This is the reason perhaps why Yi Fu Tuan's *Topophilia* (1990), which so poignantly elucidates the abiding bond between people and place, is one of my favourite reads. When I was a senior fellow at NMML, I had a chance encounter with Thích Nhất Hạnh, a Vietnamese monk, whose workshop on mindfulness I attended to great benefit. I had a *guru* who guided me about the precepts of Jainism. I think that my spiritual inclinations lessened my craving for positions and posts. I opted for meditative ways to work at one place.

By this time, I had also developed an increasing interest in the merits of meditation. I had attended programmes and lectures and decided to settle for *Vipassana*, a Buddhist practice of meditation. I came to realize that meditation is experiential. It can neither be explained nor does it have one universal outcome or impact. Even as a practitioner each time I meditate, I face a new dimension of challenges and responses. I seem to have translated this elusiveness so inherent in spiritualism to coin a new and my own definition of place. I think you probably realize by now that I want to redefine concepts and create new words when the prevailing vocabulary fails to express what I mean to communicate.

To me place is not just a creation of latitude and longitude, site and situation, land and landscape, temperature and rainfall, soil and vegetation, skills and techniques, customs and traditions. Beyond these visible, measurable, quantifiable and qualifiable traits, characteristics and elements there is something more that makes for a place. What is that? At this point of my understanding all I can say is that it is "the chemistry or essence of all these elements which creates an inexplicable yet intangible energy or force which I am tempted to liken with *atma* of place. Like soul to body, this resonates place, explains place and makes place a condition of possibilities and infinite potential. Because of such an intrinsic and fundamental nature, place has the inherent capacity to situate, hold and perform different roles."

I did not have to look far and wide to search for the role of place. All I needed to do was see close and clear. It is a tradition that the tea served in our home arrives in a dainty porcelain kettle with a tea cosy, kept firm and tight until the leaves have brewed. The brand of tea remains unchanged: a mix of Assam with Darjeeling. I have sipped this every morning and evening for a lifetime, yet it took me nearly 60 years to be attentive to the text printed on the carton of its package. Here is what it reads: "Darjeeling is a *tea* grown in the *Darjeeling* district, Kalimpong District in West Bengal, India, and …" Sometimes the common goes unnoticed, familiar appears unimportant and the ubiquitous is taken for granted. It is because of this opacity that we miss out much in life and diminish the quality of our lives.

This is the case with place. It is present everywhere, it manifests itself in a variety of ways, yet its role goes unnoticed and unappreciated. It takes mindfulness to realize that there are products and goods that owe their specialty and allegiance to place.

After the cup of tea I moved around to discover what else carried the sense of place in my home. Pashminas were what I found in my mother's cupboard. These heirlooms are featherweight, ever so soft and intricately embroidered. Each winter a Kashmiri vendor comes to our home with an urge to sell pashminas to the family. He often repeats the sentence "…only in Kashmir can be made such a soft shawl, the drape is washed in the waters of the River Jhelum and the wool is of the goats which graze on the soft alpine meadows in the highest altitudes of the Himalayas." In a bid to fetch a higher price he adds that this is a unique piece and will not be available again!

I looked around my home and located other such items of interest: the basket of fruit carried Alphonso and Malihabadi mangoes, in the launderette hung Solapur terry towels, and the *mandir* (prayer room) had the Mysore sandalwood incense.

I chose to create a new phrase and decided that such goods which carry geography within their ambit were to be called "Place Goods." This was a departure from the official or legal expressions such as protected designation of origin, appellation of source, indication of source and geographical indication goods. I felt that the term "place goods" would impart a direct allegiance, meaningful association and strong affinity with a place. The usage of the word, "place" preceding the word "goods" and not vice versa was my measured decision. It is place that makes possible the goods and not the other way round though the role of the goods in rendering eminence to place cannot be overlooked.

I scouted to find out where and how place goods acquire a legal recognition. I boarded a flight from Delhi to Chennai and headed in a hurry to the office of the Geographical Indications Registry, Chennai. The officer at the window told me that while all applications and registrations for goods seeking geographical indications were admitted in this department, the information was not readily available. Undeterred, I pipelined hundreds of Right to Information (RTI) applications to the office. Each time a response arrived I would open it like an excited child with a gift box. These arrived in large numbers and in the end I had all 199 of them. I found in my kitty not just Darjeeling tea, Kashmiri pashmina, Basmati rice but the likes of Monsooned Malabar coffee, Vizhakulum pineapples, Punjabi phulkari, Chanderi sarees, Gadwal sarees and many, many more. The RTIs provided me with a wealth of information on origin-based goods. The lens of looking at the RTI material from the perspective of theory of place seemed ever so exciting. It is almost as if the material were crying out for such an approach, analysis and recognition. These products and goods became a vivid presence of *Made Only in India,* by virtue of this country being their exclusive place of origin.

Made only in India

I realized India's place goods are a juxtaposition of all kinds and shades. Within the ambit fell places for weaving sarees, decorating metals, moulding clay, casting idols, picking fruits and flowers, extracting oils, stringing beads, knitting baskets, producing food and many more. The region of the production of these place goods could be as small as a room to as large as an entire river valley or even spilling across neighbouring states. To be honest I had not known or even heard of most of these place goods.

My thoughts were mixed: how rich, diverse and creative the people of India are and how little I know my country. Was I also discovering a new theme for research?

My imagination leaped as I tried to comprehend both the goods and the place. Here is one I enjoyed: Nalli is a chain store of sarees and fabrics in India. The collection displayed at their showroom carries labels like Chanderi, Coimbatore, Kanchipuram, Mysore, Nangavalli, Odisha, Pochampally and Valkalam. Each of these is the name of a fabric and also of a place. It occurred to me as an interesting fact that a confluence of rootedness and tradition to place has given the saree a dominant share of 37 out of the 199 geographical indications goods registered at Chennai. My wardrobe does not have a single saree, but I simply marvelled to learn that there are 600 ways to drape this traditional Indian dress.

While the goods took me places, to trace the antecedents of the idea of creating a law for place led me to read John Locke (1690), who observed that "in the state of nature, every person has… a property in his own person, nobody has any right but to himself. The labour of his body and the work of his hands we may say are properly his." I was equally impressed by Jeremy Bentham's utilitarian theory. He stated, "let us argue that since the cost of imitation is always lower than the cost for development, imitators can successfully drive inventors out from the market, if inventions are not protected in some form or the other" (1789). At an international level the subject has engaged economists, sociologists, political scientists, legal professionals and those involved in negotiations relating to intellectual property rights in the domain of trade. The fact remains that at the fundamental level the concept of geographical indications is about the distinctiveness in the goods only because of the uniqueness of their place of origin. It was clear that it was a worthy theme to geographically explore.

The strongest votary of the protection of geographical indications was France. The lawmakers of France argued that the names Champagne and Burgundy cannot be used by manufacturers elsewhere[2] in the world because wine from other manufacturers has not been produced and brewed in Champagne or Burgundy! Tracing the origin brought forth that it was the treaty signed at the Paris Convention for the Protection of Industrial Property in 1883 which, for the first time, made a reference to geographical indications. India passed the Geographical Indications of Goods (Registration and Protection) Act, No. 48, a century later in 1999 and set up the registry in Chennai in 2000 to deal with their legal issues and registration. These goods carried a purely geographical licence and the name of the Act clearly and loudly indicated this. My book *Made Only in India* was published in 2015. No geographer had earlier written a book on this theme in India or abroad. None of the departments of geography in India had ever held a seminar or organized a conference on this topic. There was not a single article on geographical indications in any of the geographical journals of India and, to date there is no doctorate in India under this title. It does not find itself in any course or syllabus of the geography departments in India. The theme beckoned itself to me. The purpose was not to seek to describe particular places in detail or to deal with a particular product at length, rather my concern was to discern the various ways in which India is recognized and represented through these goods. Unlike *Voice of Concern* or *Vulnerable India*, in this research I was neither critiquing or debating a particular viewpoint, nor testing a formal hypothesis. In a sense it belonged to the category of a blue sky research which unfolds itself along the way.

As a first step, I was keen to know why the French had become the initiator of a law for place towards the end of the 19th century? Was the idea fuelled by the concept of the regional school as espoused by the French geographers, Paul Vidal de la Blache (1918), Jean Brunhes and Frédéric Le Play? Regional geography is ideographic, concerned with the unique and revels in the particular. Le Play developed a geographical account of France around the categories of place-work-folk and so produced the idea of cultural homeland as an area of axe and plough, which belongs to the people who have carved it out. The French tradition of *la géographie humaine* associated with Vidal de la Blache, at the end of the 19th and beginning of the 20th century, was an influential school in human geography when a law for place was being constructed. Although the word "place" was not their core concern, but rather a focus on *genre de vie*, it could capture the complex interplay of the natural and cultural worlds in parts of France. I extended this grain of argument to the fact that goods exclusive to a particular area will always be unique. Along with the regional concept and the ideographic perspective, I focus on the interplay of the basic tenets of the geographical paradigms of determinism and possibilism which underpin the relationship between place and people, and I extend this to that of produce and goods.

When I was invited for a lecture at the Institute of Rural Management Anand (IRMA), Gujarat, I visited the Gandhi ashram at Sabarmati, Ahmedabad. On a signboard within the ashram was etched Gandhi's definition of *swadeshi*: "a spirit in us which restricts us to the use and service of our immediate surroundings to the exclusion of the more remote." The immediate surrounding is a sure reference to place, be it first and foremost at a local level.

Identifying place goods and learning about their location, characteristics and types and also searching for a law of place was just the beginning of the discourse. As the research proceeded I noticed a distinct historical trajectory to the origin, glory and even demise in the case of some place goods in

India. In terms of numbers, the majority of place goods predate the arrival of Islam, another substantive group belongs to the Mughals, while barely a handful originated during the colonial period. There was not a single place good registered at Chennai which belonged to the post-1947 phase.

I wanted to trace the distinctive situations of demand and supply which created such a pattern within different historical phases. The challenge was to have a holistic understanding of each of the place goods and then draw a collective narrative. The sources of information were widely varied: handicrafts took me to the National Institute of Design, Ahmedabad, mangoes to the Indian Council of Agricultural Research Library, Pusa Institute and miniature paintings to the Indira Gandhi National Commission of Arts, the National Museum of India, and the National Gallery of Modern Art. The variety of place goods meant that there was no one shop or website which could meet my demands. Exhibitions, craft *melas*, museums, catalogues, travel magazines, labels on goods of all kinds in shops of all variety – there was no end to my hunting and gathering. Even when I went to visit my family in England, I set aside a few days for the Victoria and Albert Museum, London. The latter has an exclusive section called India House which carries a range of goods that had been gifted or were taken by the British during their colonial rule. I carried a hunch that there would be many more such place goods on display in the museum. I was proved right.

I clearly wanted to see, taste, touch, smell, feel and hear the place goods. As a geographer this was a totally different experience of a field in which to work. For me, this was a new and different type of fieldwork.

When kings were patrons of temples, they believed that offering the best of goods and produce was one way to serve, please and reach divinity. The temple was thus the driving force for the destination and destiny of place goods. In contrast, the *darbar* (court) was the chosen destination for the production of place goods in Mughal times. To keep vigil on the production *karkhanas* (ateliers) were attached to the royal palace. Here goods were produced not for divinity but for nobility. During the colonial period, apart from the plantations that were nurtured and today claim an origin tag, many of the existing place goods were systematically destroyed.

To inquire into the destruction of place goods I decided to focus on textiles for three reasons: first, the destruction of textiles from India is well documented; second, textiles formed the backbone of the Indian economy from the 19th to the mid-20th century; and third, textiles continue to form a sizeable share of the place goods of India today. I do not mince my words to conclude that

> *it was the* British who had taken many place goods to great heights of production and fame and they are the same who plunged them into the worst abyss. The destruction was neither innocent nor an accident. To appease their insatiable commercial interests the colonists laid out a wicked path for virtually the wholesale destruction of [goods] Made Only in India.

I learnt that the age of the mighty Hindu kingdoms, just before the onset of the Turko-Afghan invasions, is often called the apogee, the highest point in the development of our culture. The notable historian Arthur L. Basham titled his most popular book, *The Wonder That Was India: A Survey of the Culture of the Indian Sub-Continent before the Coming of the Muslims* (1954). The word "wonder" is a tribute to the glory of India once. What followed was a cascading ruin of India.

Many of the craftsmen who nurtured place goods today live on the brink of poverty. The income of weavers of carpets or sarees or those who carve idols of wood, are such a pittance that the next generations in the family are not lured to continue. This is a sure sign of an impending extinction. The polluting water and air is poisoning places and destroying agricultural produce like Assam tea, Pokkali rice, Vazkhakkulam pineapples or the Udupi jasmine. Copycat brands employing descriptors such as Lanka Darjeeling or Hamburg Darjeeling to mischievously sell these as Pure Darjeeling are a menace for this beverage. Imitations by the Chinese of the famous Banarasi sarees are creeping into the Indian market.

I distinctly recall I had taken a conscious decision to veer away from the vulnerable, homeless, voiceless, jobless and chose instead to gravitate towards the banks of India's diversity and splendour adorned with her products. I cannot deny that I found beauty, taste and finesse of India through place goods, yet the book led me to a strange sadness and sorrow. In spite of India's talent and ability to harness place to create some of the most valued goods, today most are unrecognized, many are being destroyed and none is given their due worth. Are these vulnerable? Are they prone to a disaster? Who will raise the *Voice of Concern* for them?

I do not claim that these Made Only in India goods are better – I adopt no such illusion. All I say is that these are the goods which are unique because of their allegiance to place. Similar to art and tradition they are worth valuing for their own sake.

The first chapter of the book reinforces that "Place Matters" and the final chapter carries the title "Place Gathers." Both resonate a philosophy of how important it is to honour goods which carry the *graphy* of India.

Navigating the theory of place my concern was that the book *Made Only in India* be able to chart a new lens to understand the geography of the country. A panellist at the book release function of the book said "There is Made in India, Make in India and now Anu has added a Made Only in India!" I was pleased when Routledge ordered a second print run of the book in the first year of its publication. If a mindful sip of a cup of Darjeeling could lead me to *Made Only in India*, then how did I set out to write *Mapping Place Names in India*?

Mapping place names of India

Vice Chancellor Dinesh Singh visited my home for the first time on 12 June 2013 to congratulate my parents when I received the Amartya Sen

Award. I recall the day and date not because it was my birthday but for the reason that in less than a few minutes of his presence, he named my home *Sunehera Ghar*: a golden home. Our home is neither decorated with artefacts of gold nor does it carry the glitter and glamour of the lifestyle of a celebrity. By any standards of urban living it is spacious, spartan and simple and, at 65 years old, carries the weathering of time. Professor Singh has a way with words and often creates rhymes. While the name Sunehera Ghar has endured, the moment froze an inquiry: Why did he name this place Sunehera Ghar?

My mother presented me with a copy of the Constitution of India on this particular birthday. When I began to read this elegant white and gold embossed coffee table production by the government of India, its very first article stated: "India that is Bharat, shall be a Union of States. One nation, two names." 'Interesting,' I thought to myself. On the same day, two very different sources are suggesting place names to me. Nothing in life is a coincidence. The ready acceptance of the publication of my essay titled "Value of Place Names" in the *Economic and Political Weekly* added merit. I had just signed *Made Only in India* to Routledge and my interest in the role of place was well sustained. The fusion of name with place seemed an interesting theme. Geography is about difference and a name locates and differentiates between places. It has worth.

To be honest I also thought I had survived tougher terrains when writing on the discipline of geography, ecological perils, disasters, place goods and so felt confident that the study of place names would be relatively easier to draft. By this time I was weary of crunching numbers and throwing in RTIs and I thought a name would be a simple enough and fascinating entry into the nuance of the theory of place. I am ever keen to work in lesser ploughed fields. Moreover, as my earlier writings had saddened me, I felt that selecting the theme of place name would be joyful. Here I was to deal with an innocuous abstract object. A name. A noun.

I could not have been further removed from reality. From the very beginning I faced problems and hurdles.

There was no book in Indian geography on this theme to date. The study of place names was not a part of any syllabus in the discipline. There had never been a discussion or debate or conference on this theme by any geographical association in the country. Unlike the registry at Chennai for geographical indications, there is no department in government which exclusively handles the issue of place names. Whereas a few countries world over have created a place name law, India still had not crafted one. There is one Place Names Society of India, in Mysore, Karnataka and only a few issues of its journal, *Studies in Indian Place Names*, were available at the Indian Council of Historical Research Library, New Delhi.

An additional impediment is my tendency to research into what I call the all. In *Voice*, I wanted to meet all the presidents of the NAGI; in *Paradise in Peril* I tried to cover all the land uses in Kashmir Valley; in the book *On Disasters in India*, I wanted to include all the research on disasters in

India; in *Vulnerable India* all the representative indicators for building the vulnerability index for all the districts had to be garnered; as was also for *Made Only in India*, all the registered geographical indications had to be sourced.

In the context of place names, I realized that even if I lived many lives and hired an army of research assistants, I would still not be able to research all the place names in India. A topographical sheet prepared by the Survey of India, on a scale of 1:50,000, normally picks up all the place names in the tract being represented. On average, a sheet of this kind carries around three to four thousand names. On extrapolating this for the whole of India, the estimated figure worked out to be 1.5 to 2 million place names in India! A mind-boggling and daunting number indeed. I was still determined that my research in some way or the other had to be pan India.

I set out to research the nation's names and came across India, Bharat, Hindustan, Jambudweep, among others. Why does India have a variety of names? What are the interpretations of these names? Which among these multiple names have endured? What explains their continuity? While working on such questions, I realized that a voluminous quantity of ore is to be removed before a diamond can be found. I cautiously decided to work on all the subnational units, that is, the 29 states and seven union territories of India, as in mid-2019. But I was clueless as to where to look and how to proceed with my queries!

Research, I realize is hardly a straightforward process – set the goal, outline the procedure, mobilize the resources. Most research methodology books are pedantic and offer a form of work which suits stereotype theses, projects or teamwork and rarely apply to blue sky type of research which essentially is curiosity led. So I did the next best thing; I jumped into the rough sea hoping to reach a shore, any shore for that matter. Since I had recently received a copy of the Constitution of India, my intuition guided me straight to the Library of Parliament, New Delhi.

Parliament Library, New Delhi: Compared to most other libraries in Delhi, the Parliament Library is spacious, clean and air-conditioned. A food counter at the doorstep of the library serves meals and fresh fruit juice all day long. A glass of fresh orange juice costs Rs 10 and the bill for a meal rarely crosses Rs 30. It was clearly evident that the system ensures that parliamentarians in India enjoy the fruits of subsidized food, accommodation, and transport so that they serve the country well!

The Parliament of India came into force in 1952 after the country had polled its first general elections. Up to 26 January 2016, the Lok Sabha had met 192 times over 436 sessions and the Rajya Sabha had 244 sessions. Among the hundreds of bills, a total of only 23 pertained to the names of states and union territories. Among these, 17 bills were pertinent to the names of newly formed or organized states, and six were specific to a change in their names. I accessed all 40 debates in the House of Parliament when these bills were tabled. At this stage I felt I had acquired a direction and plan, and I was no longer a rudderless ship on the high sea! As a geographer, I had discovered a new, rich and authentic source of data. I had never before read a parliamentary debate and I was keen to know how the practice of place naming could translate into a theory of place names.

Each debate runs into several pages because every word spoken by the parliamentarians is recorded verbatim. When browsing the debates, I realized that parliamentarians digress, quarrel, retort, explain, and expand about much of everything that is irrelevant, and in the process the relevant often gets buried. Many a times sifting the debates tested my patience. But finally, I managed to write the chapter titled, "Democratization of Place Names: Parliament Debates the Names of States and Union Territories."

Curiosity led me to ask: What is the durability of these names? I had learnt the sieve method of map interpretation in my undergraduate studies; I filtered the 2019 political state and union territory map with their latitudes and longitudes on a series of 40 historical maps in Schwartzberg's atlas. Through this overlay, I classified the durability of these names into four categories: (i) continuous (none fell in this category), (ii) frequent, such as Punjab (iii) discontinuous, such as Karnataka, and (iv) occasional, such as Haryana. It was enjoyable to apply what one had learnt to create more to learn.

The above experiment brought home how the stock of place names change. Then additional questions marked their presence. How and why do place names change? What are the kinds of changes in place names? Who is instrumental in bringing about changes to place names?

As I was groping with how to begin the narrative of historical change in place names, R. Ramachandran emailed me the manuscript of his book under the title *History of Hinduism in India*. There could not have been a more fortuitous situation. He provides in this book an elaborate discussion on the arrival of the Rig Vedic people and the manner in which Hinduism spread in India. I was able to quickly relate place names with the successive stages of territorial spread of the Rig Vedic people in India. This helped me finalize a chapter under the title "Sanskritization of Place Names." I was aware that the famous sociologist M.N. Srinivas had invented the term Sanskritization to refer to the process of upward social mobility of the lower castes to the upper ones. I, however, adopted the term Sanskritization in an entirely different context. My idea was to capture the process of assigning place names during a particular historical period. Here I was reminded of the Tamilian scholars whose strong pride in their own language had led them to unravel Tamil place names prior to the phase of Sanskritization.

Depending largely on the material at hand each historical phase had to be dealt in its unique way. For the impact of Muslims and concomitant Persianisation of place names, I picked names of urban places which carried a flavour of Persian or Urdu *zubaan* from the Town and Country directory of the Census of India 2011. To handpick the Persianized names from 7,935 names of towns, I enlisted the help of a friend who was adept at the language. These Persianized town names at first sight seemed a heady hotchpotch, but when I calmed to read them one by one I was able to discern a distinct pattern. Some names were religious, such as Aligarh; some glorified the name of the ruler, such as Shahjahanabad; many confirmed a victory, such as Fatehpur Sikri; a few others represented the Muslim cultures, such as Qazigund.

I was no less keen to decipher the colonial impact on place names. I had already researched the colonial role in transforming and damaging the geography of India. This was visible in the perils of Kashmir Valley, within the disasters of India, and also in the ruin of Made Only in India they caused. The fact that the British manipulated everything to suit their insatiable economic greed was not new to me. Somewhere along the line, I had tripped upon the information that W.W. Hunter, the officer who authored the *Gazetteer* of India, had communicated in 1869 to the Home Office, London, an urgent need to adopt a system of transliteration of proper names in India and had even outlined a methodology for this purpose.

I was keen to read Hunter's blueprint of orthography in original. This took me to the National Archives of India, Delhi. A stately building, imposing and impressive, this custodian of the records was earlier known as the Imperial Records Department. I knew I was at the right place, but when I forwarded my request for the report I could never have anticipated the procedural laxity. It took 60 days to get this 40-page document! After all this wait, I only quoted less than ten lines from it in my book, yet I was convinced that the first-hand learning that the original provides can never be replaced by a duplicate.

I classified British impact on place names in India into two different categories: first in the nature of Englishization, and second in the spirit of Anglicization. I know the word Englishization is not in the dictionary but I coined it because I felt it necessary. You see I had identified a marked distinction between the two processes of place names that the British had initiated and I clearly could not have put the two together into one container. Here is why:

Under the scheme of Englishization fall the names of places discovered or created by the British, such as names for tea, or coffee plantations, hill stations, railway stations or cantonments, among others. The British legitimized their claim by lending them English names. This can be illustrated by names of hill stations like Mussoorie or Dalhousie. I called this the process of Englishization of place names in India. In contrast the word Anglicization was readily available in the dictionary, and fitted the second process perfectly.

To comprehend this, one needs to understand that when the British set out to achieve their intent to commercially exploit India, they were horrified to find that place names of India came in several languages, with different pronunciations and varying spellings. A virtual headache hit them whenever they had to deal with this jungle of place names, so they set out to standardize them.

I noticed that the British effort at standardization of place names came from a global call. The first International Congress of Geographers, at its forum held at Antwerp in Belgium in 1871, discussed the universal problem of spelling geographical names. The matter was again raised at the 5th International Congress of Geographers, held at Bern in Switzerland in 1891. Here the German geographer, Albrecht Penck, launched the idea of a world map on a scale of 1:1 million (Crone, 1962). It was Penck who made the case for a global standardized writing of geographical names.

The British reports and documents clearly state that they standardized and brought order to the mayhem of place names. While this is partially true, let us not forget that to achieve this, they screwed up the spellings of names to become English in sound and character, a process the dictionary defines as Anglicization. Anglicization was a single measure which altered the complexion of place names right across India in one go. It is under such a process of change that, for example, the name Dilli was corrupted to Delhi.

To get to see how dots join global to the local, Europe to the colonies, Britain to India ... all this and more, became my leanings along the journey of engaging with geography.

But the story of place names does not end here. It could not. What did India do with place names when it gained independence? Did nationalism herald a change in place names? What was the scale of this change? Does India have a policy for reform of place names? While I endeavoured to answer each of these questions, during which I observed a tendency for place names to gravitate back to their original names, my eyes were keen to see how far into the future place names would go.

In research each has their favourite, the one dearest to me is to look into the future and wonder what will be there in the coming years. While writing *Voice of Concern*, I warn of the decline in the quality of research and teaching in the discipline of geography; in *Paradise in Peril*, I presage the emergence of an anthropogenic lake; in Ajmal Khan Park I see wholesale vandalism of the park; in *Vulnerable India* I forewarn of serious and multiple impending disasters; in *Made Only* I take the view that all these tags will be wiped clear if sufficient protection is not provided. After gaining sufficient insight into the problem I like to foresee outcomes.

In the case of place names, my main concern is the place names of the future states and union territories on the map of India. This was a challenging task as I needed an appropriate lens to telescope into the future. I drew on one axiom. Names of states and union territories will change because the spatial contours of these administrative units will change. The latter will change because the people of India will demand this change. When reading the debates on place names, I found that parliamentarians across party lines would every so often snatch any opportunity to digress and voice the need to form a new state. Their shrillness is the call of an aspirant state. This took me to three lead questions: (i) How many and which are the aspirant states? (ii) Which strategies are they working out to achieve their objective? (iii) What names have the seekers and strugglers given to their demand for new states?

The idea of aspirant states making a popular appeal in India took me to the heart of geography: the concept of regionalism. I again felt the necessity to coin the term "placenism," a new word. By this time I had learnt that new words arise in the mind when others are not available to express exactly what one wants to convey. The reason for introducing placenism to the dictionary of geography seemed urgent and here is why I felt so.

Regionalism is the bond of people with a territory. Within a region there is often subregionalism. I find that currently in India these subregionalism pockets have lifted their heads and a few are up in arms for separate identity. In this context the usage of the word "sub" seems derogatory because it implies that the majority defines the core character of the region and within it the minority has a substatus. I knock out and want to do away with the prefix "sub" for I feel that it puts the minority on an inferior or lesser rank vis-à-vis the majority. To avoid ambiguity, I invent a fresh term, placenism, to aptly capture the spirit of subregionalism. The concept of placenism rests on the premise that people are conscious and proud of their place and demand its autonomy.

I was able to count 41 placenism movements in India which are aspiring for statehood. Each is vying to salvage a place for their name on the map of India. How soon would these place names signature the map of India? Much depends on the nature of interaction between civil society, democracy and politics. Names on the map of India have been changing over the years, and in recent times with greater speed. Of the 17 names of provinces listed in 1941, seven were obliterated by 1951. Of the 29 state names listed in 1951, 11 had vanished by 1961; of the 31 state and union territories names in 1971, two had no place by 2001.

The writing on place names seemed never ending… yet I put a full stop and mailed my manuscript to Routledge in January 2018. It was published the very same year.

As researchers we are often trained to look for and focus on the uncommon, to discover the unknown, the unseen, the unfamiliar, but what attracted me was the everyday of everywhere amidst which we spend most of our lives. I spent a decade studying goods in places and names of places. They were so commonplace that geographers in India had overlooked them. These held my fascination.

Moreover through the names, as much through the goods, I was keen to understand not just the names and goods, but I was trying to flag the rooted value of place in India's culture, indigenous knowledge and skills.

Who was I writing for? I clearly was not writing textbooks, those were too straitjacketed and I felt they could wait their turn. I wrote primarily for students aspiring to be geographers and keen to know about geography *in* India and the geography *of* India.

I stagnated at the rung of an assistant professor from the age of 28 to 43 years, these 15 years meant a heavy teaching load, administrative chores and running many errands in the interests of my institution. It was only by putting in weekends and staying from 08.00 a.m. to 08.00 p.m. at DSchool that I was able to get through my writings. I was exceptionally diligent, slept few hours and reworked the drafts of my ideas persistently, insistently and almost obsessively. It seems my soul hankered for a place and anchored at a place.

The University of Delhi has been my workplace and also my LifeSpace. I stepped into its portals when I was 18 and today am 60 years. The university is a place to learn but it is also a place about which to learn.

Mapping the University of Delhi

The University of Delhi is not a small place. Its approximately 400,000 students are enrolled in its 79 colleges and 82 departments. The spatial arrangement of the university is binary with two campuses, one in North Delhi and the other in South Delhi. The former was the first to be built and remains its nerve centre. It houses the office of the vice chancellor at the Viceregal Lodge, administrative blocks and a large number of prestigious colleges. Students from Delhi and across India throng to seek education within its portals.

I raised a set of typical geographical questions rooted in the concepts of location and situation, spatio-temporal evolution and emerging patterns, catchment area and area of influence, ecology and environment, crime and safety, services and transport and those of vulnerability and resilience with reference to my university.

I am convinced that spatiality is the focus of the study of geography and no better tool can be harnessed than the map to pull this all together. There is nothing I have published without putting a map into service.

I have taught the course on cartography to both undergraduate and postgraduate students for over a decade. To me a map was always a blend of art and science where one absorbs and reflects the other. Without a command over both, it would be difficult, if not impossible, to create a map. Teaching layout was always a challenge as the positioning of the legend, choice of the thickness of lines for the borders, placing the title and subtitle, the scale and sometimes the inset, were features that all had to be combined. The difficult part really was how to teach the student the responsibility to transfer map information to the map user. To teach the value of harmony, composition and clarity to students was most challenging. Nth numbers of thematic atlases prepared by students on themes as varied as population, development, politics and environment, among others, have been signed by me. I recall endless corrections of the legend, title, scale, and most importantly my insistence on the selection of an appropriate technique for mapping any set of data. I consider a map an exceptionally effective tool for representing reality. I am proud that the map belongs to the kit bag of geographers and others have to borrow it from us. I immensely enjoyed the occasion when I delivered a lecture titled *Power of Maps* to an audience of government officials attending a workshop called Decentralized Multi-Level Planning. I liked to use maps in time series to analyse geographic processes and with GIS this sifting and sorting became all the more exciting.

The drawing of a map takes its share of time. I think it takes more to edit a map as a visual of spatiality than to edit a text. I have not hesitated to seek assistance from professional cartographers and the likes of Mohan Singh, Panjab University, Chandigarh have shown patience and dexterity with my endless drafts of maps. In my stock of publications there may be over 350 maps on diverse themes which I have enjoyed creating. As far as I recall I have not copied or lifted a single map from any prior source. It was within this geographical experience and thinking that I decided to use maps as the medium for mapping a variety of dimensions of the University of Delhi. Each map comes from a story and each map tells a story. Here are the narratives about some of my favourite maps on the University of Delhi.

Story 1 (1998): A Nostalgic Gift: The First Map on Miranda House.
Filled with nostalgia, I so wanted to create a special "gift" when I "was to leave" Miranda House in 1997 to join as faculty at the Department of Geography, DSchool. To my surprise I "discovered" that never had a map of the college been created. The origin of Miranda House dates back to 1948 and the college had offered geography since 1965. Yet until 1997 there had been no map! Where do I begin? Surely lying around somewhere there would be a drawing plan by Walter George, the architect who designed Miranda House. I was able to lay my hands on a frail and fraying tracing of the first building plan of the Miranda House from the engineering department at the University of Delhi. Subsequently the college had added new buildings, extended older ones and set out sports and other complexes. The Annual Reports of the College, its foundation stones, and photographs of inauguration ceremonies helped me to piece the history of the changing map of Miranda House.

I hired an artist, one who paints city signboards, to meticulously paint the map of Miranda House on to a large size metal board. I wanted the map to stand erect and firm on its own feet. As instructed, he painted the background grey against which the out-lines of the building of Miranda were made in red. This choice of colour was influenced by the brick structure of the college. I recall that Principal Kiran Datar displayed it at the entrance gate of the college as soon as she received it as a gift. The map serves as a guide for students, parents and others visiting the institution.

Searching the files of the engineering department for a map of Miranda House brought the discovery that there was no map for the University of Delhi either.

Story 2 (2002): University of Delhi's Folio of Maps for the Wall.
Professor Deepak Nayyar, vice chancellor of the University of Delhi for the years 2000–2005 was an economist but it was evident that he was a man who appreciated heritage and valued aesthetics. It was under his tenure that the age old Viceregal Lodge received its

first impactful facelift. He was often seen meeting the team of con-
servationists and landscape architects.

The experience of mapping Miranda had left me with the
knowledge that even the University of Delhi lacked a map. It is by
noticing the absence of something that one gets the possibility to
create new. I designed a folio with a set of nine maps representing
departments, colleges and institutes, infrastructural facilities and
the historic landmarks of the University of Delhi. The most
astounding was the one I titled "University of Delhi: Search for a
Site." It came as a surprise to learn that the office of the university
had shifted six locations before it was stabilized at the Viceregal
Lodge. While mapping this movement and shifts I wondered what
the circumstances are that displace a university and so many times
too! The folio of maps I designed was published by the University
of Delhi in 2002.The maps gained front page recognition in
national newspapers and fetched several letters of appreciation. I
cherish one that came from Professor Emeritus Geoffrey Martin,
the scholar, if you recall, who had rescued my life write-up of C.P.
Singh:

> In USA we have just lost Chauncy Harris… He was nearly 90
> years of age… He was a fine ambassador for geography, a kind
> man, and an extraordinarily capable geographer… I have appre-
> ciated the folio of maps which you designed for the University
> of Delhi. They are very fine and will surely facilitate campus
> planning and growth. The packet is also a very nice idea with
> regard to their presentation.

Deepak Nayyar was a man of detail and timely action. As soon as
the heritage conservation plan restored the Viceregal Lodge, he
asked me to convert the folio into a wall map. By now I had learnt
the merits of digital technology and instead of the metal on stand
map of the Miranda days, the wall maps were converted into digi-
tal-based images printed directly onto a wide sheet of paper.
A dozen wall maps continue to adorn the main administrative
offices of the Viceregal Lodge.

Story 3 (2003): Crime and a Map for a Sustainable Campus. Maps are
a window to the terrain. I was nominated as a member of the
Sexual Harassment Committee, constituted by University of Delhi.
This set another trail of writing and maps. The newspapers often
carry the following headlines: "Campus shocked over assault on
girl," or "Women unsafe on the Campus," or "Twenty-three persons held for violence following a woman related crime." I was
familiar with the writings of André-Michel Guerry, the French lawyer and amateur statistician who had contributed to the development of criminology as a discipline. He claimed that "the different
kinds of shading on the maps of the areas of crime would not only

enable the reader to see the facts more quickly but also to appreciate more readily the essence of comparison."

The territorial jurisdiction of the Maurice Nagar police station is spread over 8 km² and includes the North Campus of the University of Delhi. The First Information Reports at this police station are a valuable source of data. Each crime report specifies, among other things, the exact location of the crime event. These inform that the North Campus of the university recorded 1,659 crimes from 1989 to 2002. Based on maps representing these data, a research article under the title "Crime on Women at the Campus of University of Delhi," was published in 2003 in the journal, *Population Geography*, brought out by the Association of Population Geographers of India, Chandigarh. I was pleased to learn that this was the Silver Jubilee issue of the journal and a dozen of my maps would appear in four-colour print. An important component was the Sustainable Security Map, elaborating how the campus could be planned to ensure security and safety for women students. The publication was forwarded to the university authorities. A response is still awaited.

Meanwhile teaching continued and the maps opened doors to the viability of training students to adopt the University of Delhi as their field of inquiry. The master's programme at the Department of Geography includes the compulsory component of preparing a dissertation. In this regard students were trained to conduct research on a topic based on fieldwork in a particular area. A handful of my postgraduate students explored and covered diverse facets of the university, such as its land use, hostels, sports facilities, traffic issues, student problems and a comparison between the North and South campuses. One of the postgraduate students wrote a dissertation on the challenges faced by blind students at the university.

Story 4 (2012): A Tactile Map: The Blind to See their Campus. India is now home to the world's largest number of blind people. Of the 37 million people across the globe who are blind, over 15 million are from India. In 2006, the government of India announced a national policy for Persons with Disability, which provisions for their physical, economic and educational rehabilitation. The University of Delhi set up the Equal Opportunity Cell in the same year to ensure equitable access to various facilities for persons with disabilities in the field of higher education.

I often saw blind students with their probing white canes, feeling and at times falling, as they walked from the bus stop to their classrooms. A concern arose in my mind: Could one devise a map that would help the blind navigate the campus? The thought connected me with Saksham, a non-government organization that works for the visually challenged. I learnt that braille is a font and downloadable.

It was new for me to work the Picture in A Flash (PIAF) machine. Once I understood it, the rest was easy to follow. I drew the North

Campus map in bold black lines and symbols. When the printout of the map was taken on a special paper called capsule paper, the heat of the machine picked these lines and symbols to swell. The blind students' sensitive fingertips follow these raised lines to navigate their destination. The interactions for the map inspired me to enrol in a PhD thesis on disability in India. The first of its kind on this topic within geography in India.

Though I was delighted to craft the swell paper map I knew that the future for the blind is not with tactile maps but more with technology – like trying to use a GPS and satellite tracking to guide the students' travel through an environment. I am sure there are many smart apps in this arena but I also know that it will be a long struggle for students in India before they can gain the status of equal opportunity in any arena.

In the meanwhile when I was included in a select group of academia in 2014 to meet the late Honourable Shri Pranab Mukherjee, president of India and chancellor of the University of Delhi, at the Rashtrapati Bhawan, I had the opportunity to gift my tactile map series to him. Such a sharing of geographical items brings lasting joy by serving a useful purpose.

The first International Symposium on Maps and Graphics for the Visually Handicapped was held in Washington in 1983. The University of Delhi got its first tactile map in 2013. Better late than never! In certain cases, I was late.

Story 5 (2012): From Losing a tree to the Census of Trees: It was Friday 13 July 2018. I found the one lone sturdy *shisham* on the lawns that look out from the windows of my ground floor office. The position is worthy to note. The dictionary defines prostrate as a "formal lying with your body completely flat, for example because you are ill or tired." The tree fell at midnight. It was too proud to let anyone witness its fall. This *shisham* was sensitive; it made sure that the direction of its fall did not damage any wall or building.

The explanation given was that the thunder and rain which lashed the city the night before had taken its toll. The blaming of nature is an easy exit route to fake and explain such situations. My book *Vulnerable India* was replete with such finger-pointing illustrations. My observation of the series of vicissitudes that the tree had witnessed was enough to convince me that it did not fall to a rain or wind, but was killed slowly yet surely.

A former chairperson of our department in 2009 had dumped furniture and a lot of e-waste at the foot of the tree. This set the tree to rot. Most in DSchool saw me doing my reading and writing in winter under its canopy. I recall a younger colleague of our department telling me that a student in the examination script had quoted the habitat under this tree as an example of place making. Was it intentional to dump the waste at the skirt of the only *shisham* in

DSchool? One cannot be sure. The stuff to be written off could have been stacked just as well in the department's storeroom.

On an afternoon in 2011, I received an urgent phone call from a student, "Mam, come fast they are cutting your tree." The excuse offered by "they" was that the branches of the *shisham* were ruining the building of the Agro Centre Economics Research Centre and therefore needed to be pruned! Is pruning not different from mutilating and chopping? We all know *shisham* is a subspecies of the rosewood and its bounty fetches a hefty price. I reached the spot and put a stop to them. This time I was chairperson of the department and so could take an official position. It was a strong tree. It tried to hold root. The photograph I had taken in 2016 shows it in full bloom. Yet in 2018 it fell prey to human forces responsible for its doom. The gardens of the DSchool are not perfectly manicured. I like their wildness of sorts, but they need some care, some attention, some time! The loss of the grand *shisham* could and should have been avoided. I became determined to learn more about the fate of trees on the campus through a tree census at a personal level.

A limb of the Aravalli ridge serves as a boundary to the North Campus. To count the number of trees I, along with my students, selected a swathe of 500 m² at two sites of this ridge. The extrapolation of the results of the survey led me to conclude that to make way for colleges, hostels, sports grounds, residences, offices, markets, and roads approximately 75,000 trees had been sacrificed on the site of the North Campus from 1922 to 2012. The case of the *shisham* was not a lone loss. I raised a question to myself: How many trees are there on the North Campus and how many more can be planted? Calibrating Google Earth images to a scale of 0.25 m, one could locate and identify sites where more trees could be planted on campus. The outcome was a dozen maps showing the best locations for trees could be planted to make for a greener University of Delhi. I forwarded *University of Delhi: Tree Census 2012* to the authorities. To date they have not set out a green signal. Instead the air thickens with pollution … and students and colleagues often choke and some use inhalers as they make their way to class.

Story 6 (2013): A New Year Gift for the University of Delhi. Over a decade of cartographic work had left me with a rack full of maps, sketches and photographs on the University of Delhi. I had learnt that Vice Chancellor Dinesh Singh (2010–2015) was not only a mathematician but also a painter. This enthused me to create a special New Year gift. I pulled together 12 maps and 12 photographs and designed a desk calendar of the University of Delhi. The maps captured the sense of place and the photographs that of space. The calendar covered two years; 2013 and 2014. I delivered it to the office of the vice chancellor. It was a grand moment when the University of Delhi took the onus of publishing it and the Union Minister for Human Resources Development, Dr M.M. Pallam Raju, was requested to release it. As souvenir to dignitaries, scholars and foreign visitors, the desk calendar took the flavour of the university to people and places far and wide. Today the reception of the Viceregal Lodge displays these maps on its walls under my signature. Maps have been a valuable tool to reinforce my love for my alma mater.

For years I taught the course on thematic cartography and supervised student atlases yet the inspiration to create an atlas never crossed my mind. My yardstick of a masterpiece is that by Joseph E. Schwartzberg, *A Historical Atlas of South Asia* (1979), published by the University of Minnesota, USA. I was a master's degree student when Schwartzberg personally visited our department and gifted a copy to our library. Its superb content and quality and the terse text is hard to match. I could not convince myself to take on the making of an atlas, perhaps because it involved a large team and colossal funds from sponsors and government. What I enjoyed instead was curating exhibitions of maps.

It is said that a picture is worth a thousand words. A map is a picture. It is a piece of art consistent with the rules of science and produced with the help of technology. I took the opportunity to curate three map exhibitions, the first in 1997, the second in 1998 and finally 2013.

Story 7 (1997, 1998 and 2013): Curating Map Exhibitions: When Miranda House was celebrating its golden jubilee in 1997, along with Dr Uma Chakravarti, a versatile colleague from the Department of History in the college, I decided to organize an exhibition. We titled it *Reflections: Miranda at 50*. To showcase the changes at the college, Uma took the photograph route while I chose a display through maps. The support of the principal, Kiran Datar, was generous. She allowed access to all the registers for admission since the inception of the college. Those were not the days of computers, and the neatly scripted pages of these ledgers carried the names, addresses and details of students. Such information was most useful to demonstrate the growth in student numbers, catchment area of the college, the changing profile of students opting for different courses, and the changing landscape of the college.

DSchool was celebrating the golden jubilee of its establishment in 1998. The school was in a kind of hyper celebration mood because the year coincided with the awarding of the Nobel prize to one of its illustrious alumni, Professor Amartya Sen. One grand highlight of this occasion was that Sen had committed to deliver a lecture on the school's Founders Day. On this occasion, I curated an exhibition titled *Reminiscences* displaying maps along with photographs to trace the journey of our department.

The exhibition in Miranda House and DSchool were satisfying but still were small displays. My solo big break came in February 2013, when Vice Chancellor Dinesh Singh gave a voice to his unique idea of Antardhwani. Antardhwani *(inner voice)* was meant to capture the spirit of the university, through a display of its research publications, pedagogic innovations, sports achievements and cultural eminence in dance, drama and music among a host of other talents. Dinesh Singh, obviously, is a man of imagination. Never before had the university pitched its works on such a grand scale. Each college and department was given a dedicated stall to set up their presentations. I was fortunate to get my lone space that was also on a prized location on the patio of the main university hall. I titled the exhibition *Aura: University of Delhi*. The display captured the facets of the university through a series of colourful maps. Along with these I crafted souvenirs for the university, with maps on its mugs and logos on scarves, the latter became a seed for the inauguration for a souvenir shop on campus. The signatures and comments in my visitors book confirm both the footfall and the appreciation received for *Aura: University of Delhi*. It felt good.

It was obviously a good time for maps. A grand opportunity came my way when I was nominated a member of the team for the 2013 production of the *son et lumière* (sound and light) program of the University of Delhi titled *University of Delhi: A Legend*. The maps were an integral part of the narrative of the programme. The show went on for days and on popular demand was replayed as a second round in 2014.

I have been at the University of Delhi from 1979 to 2020, a lifetime of over four decades. In this duration, India has doubled its population; a dozen prime ministers have held office in succession, and the University of Delhi had seen the terms of nine vice chancellors. My personal academic and professional journey from a student to professor of geography has been no less eventful. I cannot help but write out a few of the marked changes I have witnessed and experienced about geography in India in these years gone by.

Then and now

Local to global

It was in 1971 that our department inaugurated its first journal under the title *Analytical Geography*. The journal suffered infant mortality and did not last a year. In 1979, the department registered the Association of Geographers, Delhi to foster interaction among geographers in the city. It publishes a modest newsletter on a regular basis and its membership hovers around 400 or so. In a sense the department cannot boast a robust geographical society or journal of its own, yet it has today become an anchor to two of the largest geographical forums, one in India and the other abroad. One is the National Association of Geographers, India and the other the International Geographical Union (IGU).

The National Association of Geographers, India was registered in 1978. It started out with 132 life members in 1978, the number of which increased to 244 in 1980, to 1,260 in 2000, and now stands at 2,257 in 2020.

From the general secretary's desk emanates responses to queries, publication and distribution of journal, maintenance of the website and work for the organization of meetings and forthcoming conferences.

The base for all NAGI's work since 1987 to date remains the Department of Geography, DSchool. Such positioning happened when R.P. Misra, former vice chancellor, University of Allahabad, was invited to join the department as professor and chairperson in 1987 and it was around the same time that R.P. Misra took over as general secretary of NAGI. The next secretary general of the association was B Thakur, professor of our department. The baton of general secretary came to S.C. Rai in 2011, again from our department. Since 1987 there has been clear effort to keep the management of the NAGI within the department's orbit.

The IGU was founded in Brussels in 1922 to promote the study of geography and has also found a presence in our department from time to time. Our faculty, like C.P. Singh and S.K. Aggarwal have been its members and headed commissions over the years. In 2012, R.B. Singh was elected as vice president of IGU, and subsequently as secretary general and treasurer of the IGU for the period 2018–2022.

The department remains space starved and so there was just no way could a separate room be allotted for either NAGI or IGU. As a result the entire secretarial operations were conducted from the offices of those who were in charge of these organizations. More importantly, these activities were seen as personal to the incumbent and a sense of belonging to the department was never fostered.

I was not drawn to the affairs of either NAGI or the IGU. Yet one cannot deny the impact of the two forums in the department. They attract young faculty and research students and create a space for the faculty-in-charge to raise networks as social capital. Although both keep the department active and busy, I so wish their presence would have made more meaningful contribution to knowledge creation in the discipline.

Annual to semester

To fall in step with the system being followed in world class universities, the University of Delhi moved from an annual to a semester mode in July 2011. As for the faculty, the semester mode allowed one to opt for a greater variety of courses to teach and left scope to swap semesters if and when other priorities stepped in. I often wonder if the semester system allows one to dispense with a course in shorter capsules allowing students to memorize and retain their learning for an equally short duration of time. I cannot precisely conclude on the value of a semester, but the more difficult challenge has been to adapt to technological changes that have taken over the mode and method of doing geography.

Technological change

I will share two areas of technological change which I came to experience; one pertains to our reference library and the second to our cartographic lab. Both are essential spaces for any geography department.

A sea of change is the library which carries the name Ratan Tata. The Delhi School of Economics Society received a beneficiation of Rs one lakh from the Ratan Tata Trust for its library. This is how the name Ratan Tata came to stay.

The library was a quiet spacious place until the early 1990s. The windows opened to the lawns, and high ceilings, low fans and caned wooden chairs built an ambience where a book, notebook and pencil were all we needed. Photocopy machines had not made an entry and this meant that one had to handwrite notes. To cope with our demand, the library stayed open until ten o'clock at night. In this rather modest environment we developed the ability to read speedily and sift through and note the relevant information from those reference books!

In 1986 the first photocopiers arrived. The exact reproduction, low price and ease from labour were its main attraction. The popularity of the technology was so immense that when the library staff could not cope with requests, the management of DSchool decided to allocate a separate room within its premises and outsource the work to a private vendor. There was only one rider, the rate had to be set low at 40 paise per page subsequently raised to 50 paise. So popular is this service that more students queue at the photocopier than in the stacks of the library. Students even photocopy class notes. Not only do absentees get served but I think that additionally the class notes of some teachers are at such a premium that these circulate from college to college!

The installation of duplicating machines kicked out the culture of sitting in the library. With only a few students, the shutters of the Rattan Tata Library were pulled down by five in the evening. Such a state of affairs continued well into 2010.

In 2010 suddenly the library saw such an influx of students that grousing about a severe shortage of chairs to sit in the library became common. Two factors seemed to influence this return of students. The first related to air-conditioning in the library and second was the purchase of journals in digital form. Both were gifts of technology.

An endowment of Rs 5 crore from the Ministry of Finance, government of India in 1993 enabled the library to maintain its enviable collection of journals. The library's vast collection of books, documents and journals, much of which is available on an open access basis, has been a key aid for students' research. The purchase of research journals on the digital platform was a welcome step but again not without some problems. The vendors of e-journals are businessmen, costs are steep and the moment the taps run dry subscriptions are not updated. The escalating cost of foreign journal subscriptions following the devaluation of the India rupee has added to a crunch. By now we all are totally hooked on accessing journals over a digital platform. It is a different matter that I print everything I need to read because the feel of paper, turning pages and making those scribbles in the margins remain a habit that lends its own joy. I may well be a dinosaur in this Pleistocene age!

I am resilient to hold on to what makes for simple and joyful ways, but I simply cannot afford to be resistant to technological change. I purchased my first laptop in 2000. In 2002 our department set up its Remote Sensing and GIS lab and introduced a course on Computer Aided Cartography. The days of those leaking rotary pens to trace maps were well gone and over. I needed to retool. Unlike many of today's students who have grown up in an environment suffused with such products of technology, I was tech deficit. I have to admit that I have learnt a lot about these technologies from my students who are much smarter than me! Bless them wherever they may be.

If any technological software has revolutionized geographical analysis, it is undoubtedly GIS. GIS has a powerful ability to represent geographic space in geographic inquiries.

The great strength and potential of the GIS was revealed to me while I was raising queries to identify the fire disaster zones in Delhi vis-à-vis the location of fire stations. I used this tool to great benefit when I was superimposing vulnerability indicators and pitching them vis-à-vis the disasterscape. Software has no doubt become more user friendly, but keying and organizing the data before attaching them to the files of the mapping software still entails a fair share of manual and mechanical work which is time consuming. Once the hard labour is over, producing map after map at the flick of a button becomes a playful field of unlimited experimentation. This multiplicity was just not possible with the rudimentary pen and ink. But I suppose then we used our imagination more often than we do now.

The first Indian Remote Sensing Satellite System became operational with the launch of IRS-1A on 17 March 1988 and another was launched recently on 17 January 2020 called GSAT. India today has 109 satellites in orbit. Now our students use high-resolution data from satellites for mapping urban sprawl, tracking landslides, monitoring land use change, measuring ground water depletion and in many other directions. Such a technological invasion and presence was unheard of in the 1980s. We still have not acquired the expertise to use such databases to build virtual systems which allow immersive experiences with different environments. That time will certainly come too.

Some departments of geography in India are offering special courses and even diploma and degree courses specializing in Remote Sensing and GIS.

Technology is knocking hard at our doors. Our department has installed ArcGIS 10.2 version and purchased licences for ERDAS IMAGINE 2014. A colleague in charge of the programme in my department advised that we need close to Rs 50 lakhs every two years for updates and upgrades! Sometimes I do wonder if these economic expenses bear dividends of superior research output? Perhaps the more serious task would be to guide students how to harness technology and information blasts only as a means and not be subservient to it.

The facility of the Internet and computer software has greatly expedited access to information, yet I must confess that for me the struggle to hone an idea, work an argument and reach the final manuscript remains very much the same, if not more. For me technology remains simply as a tool.

Technology did, however, enable me to develop a new hobby – that of making short films. I put together the family narratives and albums and showcase these at family get-togethers. That my films are in great demand is evident from the requests I receive from others to direct and produce the same for their events. Little do they know the hard work and time it takes to make even a 15 m film!

In the same vein students and friends in media ask me to record my lectures on YouTube and to create Internet courses. Perhaps I have just not been able to shed my traditional moorings. Yet in the same breath I will not deny that I often adopt blended learning to ensure that I have an edge, but nothing can replace personal interaction with students. I believe that this is so much more humane than the faceless technologies. I do not discount the possibility that the expanding jungle of hardware and software will change the environment and I could well be an extinct species!

Gone are the days of postcards, couriers and handwritten invitations. The Internet provides opportunities for networking that in my years of earlier learning were unimaginable. Social media has put together many geographers and expanded their scope to attend conferences, seminars and workshops in places in India and countries across the globe. I am aware of many groups which are constantly in a social media sharing mode. I simply choose to stay away.

The geography community in India at large has read some of my writings, but I am definitely not seen much in person. On rare occasions I have been confronted with the expression, "Oh, so you are Anu!" I wonder if I look different with my pashmina draped over my simple jeans and T-shirt, and hair that is now a distinct salt and pepper. Perhaps this reticence has something to do with my all-girl convent and a conservative upbringing. Gender does matter.

Women enrolment

Our department, at its early stage, displayed an exceptional parity in terms of recruitment of women. In the 1960s there were two women among a faculty of three! Today in 2020, we are three among a faculty of 12. Girls comprise 41 per cent of the total students in the postgraduate programme but their presence in different social groups remains varied. Girls dominate the general category with a sizeable 55 per cent, but among the scheduled castes/tribes and other backward castes their share drops to less than 30 per cent. Never has a girl from the physically challenged category ever opted to study geography in our department! It is obviously taking a while for the important others to catch-up.

I notice women ratios remained skewed in many arenas of geography in general. Among the 43 presidents of NAGI, to date there has only been one woman – Sudesh Nangia. This position came to a woman as late as 2006. The office of the NAGI remains dominated by men, be it the post of general secretary, treasurer or editor. Perhaps we geographers mirror the scenario at the national level; so far among 18 prime ministers of the country only one, Indira Gandhi, was a woman.

It seems that even in developed countries the situation is not so bright. Since its inception in 1904, the Association of American Geographers has seen 115 presidents; among them only 15 women have occupied this rung. Ellen Semple was the first woman to become president in 1921 and Risa Palm was the second on the list to join in 1984, a gap of more than 60 years. The remaining 13 reached this position only after 2000.

Even within the curriculum of the departments of geography, issues of gender sensitivity have been slow to enter, but I now can see that these are coming around.

The model syllabus of the University Grants Commission 2000 has a course on Geography of Gender. The new syllabus created in 2018 by the Department of Geography, DSchool offers courses like Geographies of Gender and Development in South Asia, Sexuality and Space, Gender, Space and Society in India, and Trans Geographies. I cannot say geography in India has either matured or built a substantive body of knowledge but what I can say is that it is now willing to be more exploratory, diverse and expansive and moreover is definitely keen to Indianize and shed its Western coat.

Sudden increase in student numbers

Records in the office of the Department of Geography, DSchool confirm that the number of students enrolling for post-graduation stayed fixed at 15 to 20 through the 1960s, 1970s and well into 1980. Then in 1980–1990 the numbers more than doubled to 50, finally reaching a total of 100 in 2020. From 20 students in 1950s to 100 in 2020, represents a 500 per cent increase in 60 years!

In my book *Voice of Concern*, I document that up until 1950 there were 17 departments of geography in India, the number increased to 48 in 1971 and in 2020 there were almost 150 of these. While the number of departments has increased, their spatial pattern remains unchanged. The Hindi-speaking belt, more specifically the states of Uttar Pradesh, Bihar, Madhya Pradesh, Rajasthan and Haryana, continue as the heartland of geography while southern states, such as Kerala, Tamil Nadu and Karnataka, have been less eager to incorporate the discipline of geography.

Geography is taught not only at university but also in a large number of colleges which are affiliated to the universities. It is difficult to gather data on these but the All India Survey of Higher Education published by the Ministry of Human Resource Development in 2018 records that 49,093 students took a postgraduate degree in geography in that year. This number can be considered as an impressive leap when compared with 1965–1966 when there were just 2,000 postgraduate students, two-thirds of whom were in affiliated or constituent colleges.

The rise in the number of students obtaining a doctorate in geography saw an even bigger jump. In my days, the supervisor decided how many candidates he or she could take for supervision. Our department awarded its first PhD degree in 1965. By 1988 the total was 22, by 2017 had surged to 147. The comparison is interesting. We added one doctorate each year from 1965 to 1988 but from 1988 to 2017 the annual average was five. Times are heady and 93 students presented their findings in the department research forum during 2014–2017. To cope with the demand, the department at times schedules three presentations by research students in a single day. How much justification can be given to each presentation is not difficult to guess.

An often repeated lament is that university departments have converted themselves into PhD producing factories. The All India Survey of Higher Education confirms that 1,361 students were awarded a doctorate in geography during 2017–2018. In my own department, a colleague has supervised 48 doctorates, in addition to 80 MPhils.

In 1962, of the 25 universities in India, 18 had established a full-fledged department of geography. This means that 72 per cent of the universities flagged a department of geography. Today, of the total number of 903 central and state universities in India less than 150 offer the subject; less than 20 per cent. Even though private universities are mushrooming, geography, to the best of my knowledge, is absent in most. This huge gap reflects that the subject has lost importance. While departments are few, the number of students studying for a degree in geography has seen an astonishing rise. The growth in student numbers results in a squeeze on space, too few computers, a shortage of books, and distorted teacher student ratios and all this makes it difficult to maintain a high standard of teaching. We are now so packed like sardines in a tin that I often joke that soon the teacher will have to stand outside and lecture through the windowpane!

This curve of steep growth of numbers in class has nothing to do with the attraction to geography, neither is this feature exclusive to this discipline. It is a mandate of the government of India to increase enrolment in all subjects in every university. Its compelling reason has been the need to include historically disadvantaged groups, both in education and employment.

The Constitution of India, adopted on 26 January 1950, carried a provision of reservations in public sector jobs and higher educational institutions for the scheduled caste and scheduled tribe populations, who were described as socially, economically and educationally backward. This figure was to be in proportion to their share in the total population. Under this provision the figure of reservation was stabilized over time at 15 per cent for scheduled caste and 7.5 per cent for scheduled tribe. In 1991, reservations were extended to other backward classes by 27 per cent, taking the total to 49.5 per cent. A recent government notification in 2019 provides for an additional 10 per cent reservation for the Economically Weaker Sections taking the total to 59.5 per cent. Student intake by higher education institutions has thus been enhanced by 25 per cent.

The dramatic increase in the size of the student community has created acute problems for university administration. Authorities are perpetually firefighting to organize admissions, examinations and results and to ensure that convocations are held on time. This means that more serious academic issues like setting benchmarks of quality, revising syllabus, identifying talent and encouraging innovative ideas are sidetracked. Often the increase in numbers has come with a relaxation of marks for the disadvantaged category of students, both at the time of admission and also for passing an exam. This drop in performance may seem exposed at the level of higher education, but we all concede that this is a situation created largely by a rather substandard quality of education at school level in general. This means that what feeds into my postgraduate class is glaring student disparity in terms of command over language, reading and writing skills and, above all, technology.

It has become a challenge for me to decide at what level I should pitch my lecture so as to be understood adequately by everyone. The landscape of geography has changed dramatically in the last two decades.

I have witnessed a marked change in my medium of instruction. When I set out to teach 30 years ago, classes were small and definitely more uniform. Nearly all were adept at English and this was partly because most belonged to Delhi, while those from outside hailed from schools where English was the medium of instruction. Today quotas for the allocation of seats has resulted in a tricky situation; there are students who are weak in English that come from the Hindi Heartland – the states of Uttar Pradesh, Bihar, Rajasthan and Haryana – as well as students that come from the north-east where English is the spoken and written language. I retooled with *Shabd Khosh*, the English to Hindi translation app, to ensure my vocabulary does not spill too far from the meaning I am trying to convey. The battle to wade through a diversity of language and comprehension has all but been won, but I encountered a serious problem with the written script. As it became tiresome to rework the grammar, spelling and syntax I have had to politely turn away several research students. I wonder if this has been the right approach.

Recently a colleague who often travels to universities in China told me that the Chinese have deployed English teachers on their rolls whose job is to improve and edit the language and grammar of Chinese scholars' writing so that they can enter the international world of journals. Should surrogates replace the necessity for self-reading and writing? I am not so sure.

One thing I know is that we are no longer the small class where interaction and learning are intimate. Unfortunately today I can hardly recall the names of students and they have become an anonymous roll call number in my attendance register.

Weak schooling, poor command over language and the lesser confident postures are no doubt a challenge, yet at this stage of my profession it is these students who inspire me to make my lectures bilingual and anecdotal. This cohort of students has written the most endearing emails and sent me the most sensitive messages about my teaching. At times I am surprised and often bewildered and tell myself that this India is keen to learn and yet this India knows not where and with whom to learn. Few academics in India choose to or are equipped with the compassion and intellect to teach this new student enrolment in a way that will empower them. The result is that a large number will get the degree but they may well become unemployable literates.

The government of India has apparently begun to face a difficult time with the quantity vs quality situation for which they themselves are largely responsible. They have begun to devise different methods to stem and sift the tide. To cap the problem of multiple boards, syllabuses and the different standards of education across the country, the National Eligibility Test common to both lectureship and junior research fellowship was introduced in 1990. It was obviously believed that a single test would be able to set the bar of eligibility. Only students clearing the NET were to be eligible for recruitment as assistant professors in universities and colleges. The top scorers were also awarded a junior research fellowship to pursue research. I could see that the introduction of the NET set up another scuttle of sorts; a scramble to pass the exam became a time-consuming pressure. Moreover, it has often irked me how one decides a candidate's aptitude for a doctoral programme through a national level eligibility test?

The University Grants Commission and other research funding agencies such as the ICSSR have boosted the amount for scholarships to Rs 35,000 and above. This amount was Rs 5,000 per month when I was pursing my doctorate. The grant is given without any scheme of accountability attached to it: a student could well take the entire amount and not submit his or her dissertation or take the sum and submit a worthless thesis. Both scenarios amount in the end to the same thing: the sharp trend of nosediving standards in research is worrying and a University Grants Commission circular has been distributed urging faculty to sign up projects for an evaluation of submitted doctorates. The crisis deepens. Selections too have become messy and tedious.

Faculty composition and selection

The number of faculty at our department has grown; today we are 12 compared to the three at the time of its inception in 1959. The year I joined as faculty we were seven on the rolls. We taught 16 courses in the annual mode and today in a semester format we continue to teach 16 courses, though an increase in the number of faculty has allowed the introduction of new specializations and a diversified list of optional courses. Moreover, none can deny that swelling student numbers means more work linked to assignments, presentations, dissertations and examinations.

In India, a large number of departments of geography were inaugurated between the 1960s and 1970s which means that by 2020 most, if not all, have celebrated their golden jubilee. This suggests that in most departments the old guards, the founders, are either deceased or in an advanced stage of life; it also means that in most cases their first generation students have also retired or are on the verge of doing so. The faculty in the departments of geography in 2020 presents a demographic structure far more youthful than what I encountered in 1980s. They are trying to make their mark in the discipline.

A marked change is the composition of faculty. When I was a postgraduate student in 1979, almost the entire faculty had taken a doctorate from either the USA or the UK. When I joined as faculty in 1997, the teaching fraternity had become a mixed bunch with some hailing from different universities across India. I was the lone faculty with a doctorate from DSchool. Today half the faculty has taken their degrees from within the department and what is more a sizeable number among these are research students from one faculty. Inbreeding of the worst kind seems to have taken over. We are all aware of the implications of monocultures.

Gone are the days when an application for a vacancy was scripted on a sheet of paper or handwritten on a form prescribed by the university. The reference here is not to the online submissions but the content of what the authorities are now demanding. It may seem a digression, yet I would like to share what I have understood.

Changing criteria of application for an academic position

The University Grants Commission was set up in 1956 to provide a cushion to universities in their negotiation with the government for funding. It was expected to function in such a manner as to protect the independence of the individual universities under its care, but it failed to handle many issues in a timely manner. Up until the early 1980s the scope of promotions in the university was limited, the faculty pyramid structure being broad at the base and very narrow at the top. A small number of sanctioned posts for professors and proportionately several for lecturers was the scheme of things. A sense of frustration at the lack of promotional opportunities was rampant among the majority of the college and university faculty. The seething discontent impelled nearly 5,000 odd faculty of the colleges of the University of Delhi to go on strike in 1982. Heralding a victory for the Delhi University Teachers Union, the 109 day long strike forced the UGC to introduce what is called the Merit Promotion Scheme (MPS) in the same year. This marked a critical watershed in the academic culture of not just the University of Delhi but also, for that matter, all universities across India. In this mode of promotion, the upward movement of an academic to a higher status was defined as personal, where there was no need to wait for a vacancy to fall vacant in the next order of hierarchy. Such a scheme moderated the need for creating additional posts. The government clearly was in no mood to incur the huge financial burden which new vacancies would have caused.

The rules to qualify for a merit promotion were kept simple; the candidate must have had a minimum of eight years of service in the present position, including at least four years of service at the current institution where he or she is working, and if both these criteria were fulfilled they could be promoted to the next level. This was more a time-defined and not strictly a merit determined elevation in one's career. Under this scheme assistant professors were promoted to the rank of associate professors and then to professors on the basis of length of service. Some conditions of research work were applied, but these were at best simply notional or just a matter of formality.

Later it was realized that mere years of service was not an adequate criterion, so what was added was the condition to attend two refresher courses. The latter were two or three weeks of events organized by the academic staff college of a university to refresh and update junior faculty on the latest ideas and readings in the subject. To build stringency, on the eve of the release of the recommendations of the Fifth Pay Commission, the Career Advancement Scheme (CAS) replaced the MPS in 1996. Guidelines for implementation of this scheme have changed from time to time. As of now, an assistant professor at entry point is deemed as being in Stage I. He or she has to cross successive stages to become an associate professor at Stage 4 and is elevated to the status of professor at Stage 5. Apart from acquiring the requisite Academic Performance Index (API) score, one is expected to have undertaken one orientation programme and two refresher courses before reaching Stage 4. Those opting to compete in open category, can do so at any stage but they must have acquired the requisite API score by that time and also should have participated in one orientation programme and two refresher courses. For an upward movement to the status of professor one has to be eligible for Stage 5 for which a higher API score is mandated.

How merit promotions or quantification of academic performance impacted quality of teaching and research is a field of research in itself. In *Voice of Concern* I wrote "….the assured security of promotions in situ restricted the movement of faculty from one university to another, and also thwarted the spirit of healthy competition which is an essential for quality work."

Early in my profession I decided not to opt for the MPS or career advancement and consciously chose the route of the open selection. The latter mode was subject to two conditions: one, availability of a vacancy and second I had to compete with other applicants from anywhere and face a stiffer interview. This meant that I did not have to attend those refresher courses under the rules existing at that time. I also do not recall giving lectures in any of these forums as I considered it rather unbecoming if I pretended to upgrade my own colleagues on themes and topics they have over the years been teaching. Moreover, I strongly believe that upgrading knowledge is a continuous process and cannot be done away once or twice in a lifetime in a capsule of two to three weeks. There is little doubt that in matters of such choices I lost out, both on years of seniority and monetary gains, yet what I enjoyed was the free spirit and knowledge that I have competed fairly to get to where I am. Many might see this as odd! I do not know, yet no one can deny that we need to find ways where work-based methods for promotion must become the guiding light.

The guidelines for the CAS were subjected to revision from time to time and have been modified four times in the period 2010–2016. When I was applying for the position of professor in an open category in 2011, the UGC revised its guidelines for selection of faculty. This time they devised an Academic Performance Index (API) a tool designed to quantify a candidate's credentials on the basis of a set of indicators, each assigned a numerical score. The online form for application classifies the information it seeks into heads like research publications, years of teaching experience, administrative responsibilities shouldered, among a host of other parameters. As an illustration, I list below the scores assigned to different parameters in the category titled Research Performance:

Publication type	Score
Self-authored book published by an international publisher	12
Self-authored book published by a national publisher	10
Edited book published by an international publisher	10
Edited book published by a national publisher	8
Research paper published in a peer-reviewed UGC listed journal	10
Chapter in an edited book	5
Research supervision of a PhD thesis (degree awarded)	10
Research supervision of a PhD thesis (submitted)	5
MPhil/postgraduate dissertation (degree awarded)	2
International research fellowship awarded	7
National research fellowship awarded	5
Award of honour	5

Source: UGC Notification dated 18 July 2018.

I find it hard to discern the logic behind ascribing these scores. Herein a self-authored book, a research paper and supervision of a PhD thesis carry equal weight. A chapter in an edited book and supervision of a submitted PhD thesis are placed on the same footing. An award of honour fetches a paltry score of five. It is not difficult to comprehend the gross distortions that are likely to intrude while assessing the merit of candidates by applying such a system of scoring. A look at the mere categories and the weighting assigned in the API score confirms how it will fail to nourish academic scholarship and excellence. The API scheme is an inadequate measurement of quality.

The current emphasis on accountability seems to stem from a perception that until this API system was introduced, an academic had a cosy life. While some may have underused their time and misused their freedom, surely to grind all in the same wheel will crush those wanting to pursue thoughts more engagingly than converting every one of their even minuscule activities into points and decimals. I cringe at the API scores for I do not believe that everything which can be counted counts and everything which counts can be counted.

Take, for example, the game of number of publications. It is here that I have often wondered why the creators of the idea of plagiarism have not brought in a law against self-plagiarism. In the case of the latter: one is repeating one's own facts, findings, maps, in different outlets of research: so you publish a book, from this you publish its chapters in other books, then publish articles from this book, and then redo the book and call it updated. The sequence could be the other way: publish a few articles, then collate these into a book, then distribute these as contributions to others' books. It results in duplicate and redundant publications. Moreover self-plagiarism misleads readers by presenting old work as completely new and original. While upcycling and recycling have virtues in many areas of life, yet how the auditors of the API score discern the shenanigans remains a mystery to me. Besides, one could ask, is it justified that academia be judged on a numerical count of how many and where?

Personally, I am glad to be on the other side of the fence but I do feel strongly for the junior faculty who will face the wrath of the API. In the race to API, I see colleagues rush to attend or arrange seminars/conferences, some of little value to their interest. They attend just because they are scoring items on their performance chart. So desperate is the situation that to ensure that fruits of attendance add to the weight to their score, faculty are meeting travel expenses from their own pockets. In our department, there is hardly a single week when one or the other faculty is not travelling to attend a seminar.

My curriculum vitae distinctly lacks global travel or a presence at conferences and seminars. I did attend a few seminars but I changed my mind rather quickly and surely. Strange as it seems, even to myself, I have never hosted a conference. It is not that I do not see their worth, but somehow along the tides of life I began to feel oppressed with the urgent. To me the following were energy vampires: begging for sponsors and funds, chasing administration for permissions to attend, harrowing deadlines for submission of abstracts and papers for conference proceedings, banners, souvenirs, name tags, bags and even managing event managers… Moreover, against the many geographical ideas that awaited my attention I seem always to be starved of time. I shied away from invitees as I felt it would be most inappropriate to sit on the dais of a stage of hard work by others and not ever make the effort to reciprocate.

For similar reasons I never looked for opportunities to either research or teach at any foreign university. I have a large family and friends in the best of positions and countries all over the world, yet I wanted to learn in India, teach in India, research on India and also live in India. My passport endorses visas for the US, the UK, Australia, Egypt, Dubai and Nepal but it can boast of not much more. The cruise on the River Nile was the most fascinating. The beauty of the total desert, the barakhans and its luminous limestone scenery and deep blue sky fascinated me; I learnt a lot about Egyptian civilization but I learnt more about the similarities and dissimilarities between the Egyptian and Indian civilizations. I would have enjoyed travelling and seeing more of the world but somehow these plans kept getting postponed.

Despite inventing the strangest of rules, regulations, criteria, indicators and scores there remains a general dissatisfaction with selections. I am rarely invited to be a member of a selection committee which limits my sample of first-hand experience. But I am told by both colleagues and aspirant applicants that many selection committees turn into virtual battle-grounds where every kind of personal and sectional rivalry, whether to do with political background, caste affiliation, and favourites of supervisors or the chairperson, are fought tooth and nail. I wonder if the difficulties I faced to gain a permanent tenure in the colleges of University of Delhi in 1990s, was because a similar situation was in place. Have things really changed or has there been a dramatic increase in the scale and intensity of the same tendency? What I do know is that such methods have been a major depressant of academic standards and performance. We are already paying a price for what I am inclined to call academic sabotage. It is frustrating for deserving candidates and demoralizing for those upright committee members.

There are other dimensions which also remain stubbornly the same. Since the inception of geography in India in the mid-1920s, geographers in various forums, publications, and presidential and keynote addresses have repeatedly raised the following issues: a pressing need for textbooks, a call for regional texts, a requirement for more physical geographers, a need for an indigenous interpretation of geographical problems in the country, the need to radically improve the quality of research and the urgency to expand the arenas of employability for the subject. A hundred years later in 2020 the issues remain the very same.

Many things have gone wrong with the universities but this does not in any way mean that the wrongs have to be repeated and neither does it mean that new wrongs have to be committed. Nothing is cast in stone, there is no ready-made instant recipe for this malady and I can only hope that we move with determined sensitivity to overcome these crises before it's too late. Clamping and clawing down the best is hardly the way to nurture leaders. There is no dearth of talented people in India, those with creative ideas and moral integrity that can put the system on a better track. All we need to do is to identify them, place them in positions of responsibility and allow them freedom backed with the necessary support.

Meanwhile

In the meanwhile, I continue to live in a home which breathes history. It was raised on land purchased by my grandfather when they were displaced from Pakistan in 1947 and was built in 1950. The high ceilings, *roshandans*, the *angan* and the entrenched memories of childhood and of family functions bond me firmly with this place. When my family was under threat of being dislodged, I filed a suit in the Delhi High Court and spent hours using every skill I could harness, including drawing maps, to prevent our home from being converted into a commercial complex. I am losing time each day gone and yet I am keeping faith that truth will hold ground and I will not be uprooted from my roots.

It would have been strange indeed if my later works did not show traces of those childhood experiences. As part and parcel of the traditional upbringing, my home resounds with many gatherings of festivities, dinners and discussions. I started a book club and hold occasional music *mehfils* too. All these are energy refills in my life of academics which is soulfully alone. My parents are generous and it is their large-heartedness that allows these activities to continue in joyful ways.

My family culture of disciplined work helped me enormously in my own routine. My families work habits were my introduction to intellectual discipline. It is from them that I learnt that projects can and need to be completed. If you put your head down and ignore everything else this is possible. I worked my research in the ethos of such a culture.

I chose to write from my own research and I chose different themes, but never forgetting the geography surrounding my home, my workplace, and my country, India. All my research is within the frame of India. Within its latitude and longitude, I ascended the scales of study rather swiftly and quickly. I like to apply the craft of geography at different scales. From the study of a micro-space like a park opposite my house or at the University of Delhi, I moved to mesoscales to research the city of Delhi as well as the Kashmir Valley. I was most enthused when I researched pan India. The works on the status of geography, vulnerability, disasters, place goods and place names are at a country scale. This is because the most revealing patterns of distinctiveness, diversity and disparity emerged at this macroscale. I moved from local to regional and then to national but did not aspire to work beyond the boundaries of my country. Perhaps for me India was my world. I was content.

Much like India, my life too has been one of extreme contrasts. While I researched on the vulnerable, disasters and the plight of goods made only in India, I have also attended destination weddings on world class islands like Santorini, Greece and cruised down the Nile in Egypt for birthday celebrations. I hail from a Punjabi family where the norms of joint family and traditional knots are strong but I also saw among my students the arrival of the Romeo Dating Sites and discussions on the space for the gay. I switch with ease and fluidity knowing that ecotones are spaces of the greatest diversity where life blooms at its best!

The journey of any teacher is filled with more stories of learning from students than the other way round. My case is no different. Many of the ideas of my writings emerged from questions and concerns raised by students. It is from them that I learnt and for them that I wrote. Some of my bright research students went to pursue their doctorates in foreign universities and I get news from time to time that they hold positions in leading departments of geography. Nothing pleases me more. The acknowledgements in my books are filled with names of students who have helped at different times in different ways. The number of my research students has reduced with time; perhaps I was too preoccupied with my own ideas? In retrospect I think that I probably wanted them to commit more wholeheartedly to derive their own insights. Times have changed and students have limitations and pressures which take their mind away from research.

Internally the map of spatial organization of my department has undergone a change, the present computer lab was once the chairperson's office, the office today was an empty veranda space and, with a batch of 22 students, the measurement of the classroom was small. The ethos has also changed. The chairperson's room was a kind of sacrosanct place where we as students were always welcome but often shy to enter. Today students walk into our offices with their coffee mugs and cell phones full ablaze. I have been invited to many freshers and farewell parties hosted by students. There is a clear change in not just dress, language and food but with their music, dance, poetry, and photographs and what they bring to light in these programmes is that young India has got talent!

What will they do with geography? What are the changes in their aspirations? Which strategies to that end are they devising? In the period right up to the end of the 20th century the majority of students studying geography did so with a view to becoming high school teachers but today the students studying at under- and post-graduation levels are not keen to plough the school system. Their employment aspirations are spreading to arenas of technology, planning, tourism, environment, disaster management and the application of Remote Sensing and GIS techniques. The strongest is the attraction for a job in public sector. It is quite common these days that many postgraduates are not keen to start with a job and would rather stay at home to prepare for the civil services exams. Recently I was invited to a talk to discuss the future of geography in India. To prepare for this I conducted a sample with 111 students of my department to inquire about their future plans after completion of their post-graduation. Two-fifths of them were focusing on the civil services exam, one-third were preparing for the UGC NET exam for possible entry into the MPhil and PhD programme, and the remainder were seeking out work with non-government organizations or awaiting placement in the field of planning. I will not be far from the truth if I said that quite many enrolling for a research degree, even after securing a scholarship, take it as a safety net in case they are not successful in securing a government job, especially via the civil services exam. Research for the sake of research seems not to find a place in the scheme of most of them.

My being set the limits to my doing. Three facets checked my number of publications. First I remain a lone geographer. Somehow I do not value collective authorship or collaboration; my fierce autonomy would have made it difficult had I even attempted to walk this path. This obviously meant that the number of publications I could garner were limited. Second, as a geographer I was selective, I was engaging with something, not anything and certainly not everything. I was keen to expand the bandwidth of my learning. This meant that I needed to spend sufficient time researching and thinking through the field. This approach of horizontality and verticality put a limit on my productivity. Third, writing my own books remains my natural choice; I enjoy the autonomy, space and freedom it provides. It cannot be denied that a book carries its own weight and a long wait on the mind. In the end I could not write more books than the fingers on my hands.

My books, particularly *Voice of Concern* and *Vulnerable India* have been brought into syllabus across many departments of India. My recent books *Made Only in India* and *Place Names* have not received the same attention largely because they are themes which are not part and parcel of any undergraduate or postgraduate curriculums in India. I hope with time they are. Scholars in the field have been prompt and generous to take time out to review my writings. These have been published in leading international and national journals such as *The Professional Geographer, Economic and Political Weekly, Indian Social Science Review, The Book Review, Transactions, Institute of Indian Geographers, Indian Journal of Regional Science, Annals of the National Association of Geographers, India, International Journal of Mass Emergencies and Disasters, Current Science, Asia Pacific Viewpoint*, among others. The extremely positive responses of the reviewers are a well-earned asset to any scholar. For me the 80 odd reviews I have received have been an important source of inspiration and learning. I wish I had been more prolific, I still have many ideas, but I do not know how time will run now.

2020: a double lockdown

Nothing in my whole lived life could have seen a greater upheaval than that which has emerged in place within a span of the last ten months!

In January 2020, when I set out to write my *LifeSpace*, I could hear students shouting slogans against atrocities on their fellow students across different parts of the country on the issue of caste and religious discrimination and later protesting against a hike in their academic fees and hostel charges. I could see the streets and roundabouts on our university campus thronged with hundreds of students on a candlelit march, with posters and graffiti filled with anger and frustration.

On 24 March 2020, Prime Minister Narendra Modi appeared on television and announced an immediate nationwide lockdown. Covid-19 had arrived in India. Classes and examinations were cancelled, hostels vacated and students panicked to reach their homes in different parts of the country.

Within a matter of a few days, the University of Delhi acquired the character of a ghost locality. No one in sight. A virus had changed the geographical character of place, from sloganeering to pin-drop silence.

I opted for a personal lockdown where I engaged with no one but myself. I decided to write *My LifeSpace as a Geographer*.

In these times when the world is battling a pandemic, what I see is a lot of geography; theories of origin, spread and diffusion of the Covid virus, brakes on travel and tourism, spatial manifestation of new kinds of clusters, zones and regions, the exhortations of social distancing, the spread of quarantine fear and above all the crass insensitivity to the vulnerable. The pandemic raises fundamental questions of location in terms of place, space, region, area and distance. Location is the measure of all things geographical.

What will the future hold? A world that looks inward or outward, deeper or wider, global or local? The downtime will shape a new geography of places and pose new challenges to geographers.

Time stops for no one. My twin sister's children are married and she has become a grandmother, my brother's two sons have completed their university degrees. My siblings, their children and their grandchildren are settled in London. My father has wound up his export business, and my mother, after closing down her garment boutique shop, spends time reading and painting.

What has been my life as a geographer in India? There have been no collaborations or meetings with international geographers. There are no foreign universities to which I have applied to study or lecture. I have not co-authored with any geographer be it in India or abroad. I have not organized a conference in India or overseas. I have not applied for government projects or sought public funds. I have not submitted articles to international refereed journals. I have not sought or fought to be elected as a president of societies or organizations.

I have enjoyed every moment spent teaching students in class. I have enjoyed creating each map. I have enjoyed writing each word of my thoughts.

I have worked at my own pace with the drumbeat of my inner space. I do not know if these have been my considered choices or dictates of my destiny, but what I do know is that the journey to learn about this earth and its people has been most fascinating.

Life mixes joys with sorrows and one is rarely free from regrets. In my passion for geography I forgot to take holidays, spend long hours with my family, and even make my own family. Has geography been worth it? I do not know, but what I do know is that it has not been all that worthless either.

It is what it is.

Notes

1 Singh, "Image of Geography in India," 172–197.
2 https://en.wikipedia.org/wiki/Champagne#cite_note-Oxford_
   pp._150%E2%80%93153-1 and https://en.wikipedia.org/wiki/Burgundy_wine

Bibliography

Bentham, Jeremy. *An Introduction to the Principles of Morals and Legislation.* London: T. Payne, 1789.
Crang, Mike. "Cultural geography." *The Dictionary of Human Geography* (5th ed.) edited by Derek Gregory, Ron Johnston, Geraldine Pratt, Michael J. Watts and Sarah Whatmore, 129–133. Wiley-Blackwell, 2009.
Crone, G.R. "The future of the international million map of the world." *The Geographical Journal* 128, no. 1 (1962): 36–38. doi:10.2307/1794110.
Kapur, Anu. "Insensitive India: Attitudes towards disaster prevention and management." *Economic and Political Weekly* 40, no. 42 (2005): 4551–4560.
Landy, Fredric. "Geography in France and India." *Economic and Political Weekly* 23 (2005): 1658 and 1792.
Locke, John. *The Second Treatise of Government.* London: Awnsham Churchill, 1690.
Nietzsche, Friedrich. *Human, All Too Human: A Book for Free Spirits.* 1878. Edited by R. J. Hollingdale. Cambridge, UK: Cambridge University Press, 1996.
Singh, Ravi S. "Image of geography in India." In *Indian Geography: Perspectives, Concerns and Issues*, edited by R.S. Singh, Jaipur: Rawat Publications, 2009.
Tuan, Yi Fu. *Topophilia: A Study of Environmental Perception, Attitudes, and Values.* New York, NY: Columbia University Press, 1990.
Vidal de la Blache, Paul. *Principles of Human Geography.* 1918. Edited by Emmanuel de Martonne. Translated by M.T. Bingham. London: Constable Publishers, 1926.
Wren, Percival Christopher, and Martin, Henry. *High School Grammar and English Composition.* Bombay: Maneckji Cooper Education Trust, 1936.
Young, George Malcolm, ed., *Speeches by Lord Macaulay with his Minute on Indian Education,* 1935. http://www.columbia.edu/itc/mealac/pritchett/00generallinks/macaulay/txt_minute_education_1835.html

Films

Shaheed, DVD. Directed by S. Ram Sharma. Produced by Kewal Kashyap. New Bombay: Digital Entertainment, 1965.
Upkar. Directed by Manoj Kumar. Produced by Harkishen R. Mirchandani & R. N. Goswami. Vishal Pictures, 1967.
Do Bigha Zamin. Directed and produced by Bimal Roy. Bimal Roy Productions, 1953.
Roti, Kapda aur Makaan. Directed and produced by Manoj Kumar. Chandivali Studio, 1974.

Part II
LifeScapes

2 Unfinished words and untold works of an Indian geographer

S.G. Burman

Professor Savitri Gauba Burman is not with us anymore. Her association with geography and the Delhi University community of geographers extending over several decades will be remembered by her students and their students. Yet contribution by any scholar invites debate and some controversy.

Two opposing views confront the debate over Professor Burman's contribution to Indian geography: first, that she left many unfinished words and published less than what was expected of her and therefore could not create a mark on Indian geography. Second, her scholarship embraced varied dimensions of geography presenting a holistic endowment to the discipline. With her demise on the afternoon of 2 December 1995, Professor Burman is unaffected by the debate. Yet an objective analysis of the controversy becomes imperative for a number of reasons. By unfolding the facts, the analysis helps sift the validity of such views. This is important before drawing any conclusions on the works of an academician. But perhaps a more important reason is that such an analysis is helpful to present and future geographers to help define their goals within the scientific community: some which are neglected and others favoured. Such an analysis thus provides the path a geographer could take to uplift the standards of scholarship in this discipline.

Personal letters, official correspondence, notes, diaries, the collection of books and photographs, and slides were assembled to gain insight into the academic life of this Indian geographer. Discussions with family, colleagues and students formed an important part of the inquiry. What perhaps could be argued as the strength of this paper is my experiences as a student under her tutelage for a decade and a half.

Unfinished words

In her tenure as a teacher for 45 years Professor Burman officially lectured for no less than 20,000 hours and taught over 1,500 students. Even within the grim walls of the Jessa Ram Hospital when an abscess choked her lungs in October 1995, she displayed enormous dexterity at imparting knowledge. Her conversations of anecdotes, experiences, and observations had a common underlying purpose: inquiry or explanation. The quest for learning replenished her reservoir of discussion and was perhaps one of the reasons which left so many words unfinished. On the flip side, this inability to check an innate passion for learning prevented many a thought from crystalizing on paper for final publication. Many half-written notes or drafts of articles, reviews and manuscripts have been unearthed from Professor Burman's academic treasury. "A note on the geographical aspects of demographic data", "Natural resources and population growth: an ecological approach", "The ecosystem approach to geography", and "Geography and the teaching of environment", among others are proof to the state of unfinished words. A footnote or corner note in Professor Burman's own neat small handwriting on most of these papers says: "needs revision, improvement, straightening or reframing." These one word notes reflect the critical dissatisfaction Professor Burman nurtured about many a scientific explanation she penned. But then, is that not what drives a true researcher to search, an explorer to seek or an artist to create? Nothing illustrates better the intellectual curiosity and the reason for her unfinished words than Professor Burman's design for the study of the Himalayas.

Pushed up by the colliding Asian continental plates the magnificence of the Himalayas held a lifetime of fascination for Professor Burman, but she was no idle spectator. This landmass of parallel ranges, slope declivities and encapsulated valleys was still a mystery and little studied when Professor Burman joined the Planning Commission as a researcher with the Natural Resource Division in 1952. The early 1950s were the days when the Planning Commission under the stewardship of Pandit Nehru was busy drafting the First Five Year Plan. Assessing the country's natural resources formed an essential prerequisite for the formulation of these plans. Professor Burman's contribution was her, "Note on the development possibilities in the Western Himalayan Region" which was mimeographed for circulation within the Commission. At about 30 years of age with a PhD from Clark University, USA the note on the Himalayas was Professor Burman's first assignment as a researcher with the Government of India. Though employed for only a brief eight months with the Planning Commission, Professor Burman's scholarly contributions attracted many admirers. It is her colleague from the Commission, Mr Tarlok Singh[1] who involved Professor Burman 28 years later in the Himalayan studies for the Committee for Studies in Cooperation and Development. Popularly known by the acronym CSCD – with its secretariat at Colombo – it has today emerged as a recognized spokesman within the South Asian Regional Cooperation (SAARC). The hallmark of this ascent from a research fellow to a key resource person for a supra-regional-cooperation body is that it occurred without a published treatise on the Himalayas by Professor Burman. Many geographers who have published more have accomplished less. In fact, most of her staunch professional admirers are of the belief that she commanded recognition even outside the walls of the geographical community and this is praise for her established scholarship. However, the fact that she did not publish enough is a common lament of even her most affectionate critics. Here lies the need to uncover the design a scholar adopts for any study.

Design for the study of the Himalayas

Stretching curvilinearly in a longitudinal extent of 74°E to 95°E the Himalayas are vast and diverse. Only a genius or a fool would attempt to write hastily about it. While many mistake themselves for belonging to the former category and attempt books, Professor Burman could foresee that an understanding of the Himalayas is not a matter of sudden revelation but of gradual unveiling through research. Her intellectual integrity surfaces from her design to devote a lifetime to the geographical study of the Himalayas.

To a fleeting onlooker the amassed data, maps, photographs, slides, reports, thesis and notes which Professor Burman stockpiled on the Himalayas may seem an incoherent accumulation of information. But a deeper scrutiny and classification unfolds a series of evolutionary stages in her study design.

Stage 1: reconnaissance – 1952–1963

Reminiscing down memory lane Professor Burman, with her gifted ability for dates, once recalled that she first saw the Himalayas on a visit to Parachinar in 1935. Set within the Hindukush, Parachinar was a settlement around 50 km to the north of Peshawar. Three years later she travelled to the other end of the Himalayas to visit her aunt on the tea estates of Sikkim. But the genesis of Professor Burman's attraction to the Himalayas as a region for study goes back to the early 1950s although a romantic attraction was not the basis of her choice. Not flirtatious by temperament, she wanted to make a serious commitment after a thorough reconnaissance. Thus through the 1950s she travelled widely across India. She visited the Chambal ravines in early 1952 and toured the Orissa highlands, Cape Comorin, the Kerala backwaters, Nilgiri Hills and the Mysore Plateau in mid-1952. Sandwiched between these travels were the months at the Planning Commission. An assignment as a reader with Rangoon University (1953–1955) took Professor Burman through various parts of the Irrawaddy Valley and the Arakan Yomas. The latter mountains were created from the Burmese geosyncline which is the north-south extension of the Himalayan geosyncline. Professor Burman developed an interest in the land and people of the North-East Frontier Agency (NEFA) which had its roots in the two years she spent in Burma. In fact, NEFA is known to be the homeland of several tribes, which are ethnically Indo-Mongoloid but linguistically Tibeto-Burmese.

On her return from Burma there was a gap of three years before Professor Burman obtained suitable employment in India, but these years were not spent in vain. Carefully picking areas to visit she embarked on a reconnaissance of the Rajasthan Desert and parts of the Western Gangetic plain in 1956 and the coast and Western Ghats of India in 1957. A stay of two years at Pondicherry gave her a reason to visit the Carnatic region and took her to the Cardamom Hills and the Cauvery delta.

Without a day's employment with any college or university department in post-independent India, Professor Burman had, by the age of 36, accomplished a massive reconnaissance across a vast area of land which lasted for almost a decade. As with most issues in life it is difficult to discern if choice, chance, or a preordained plan operates to fix our areas of work. But one aspect is clear: at the end of the 1950s the geography of the Himalayas was little studied. Much changed in years ahead, and while many geographers leap-frogged from one area to another Professor Burman maintained a steadfast commitment to her study of the Himalayas. They were a challenge which beckoned Professor Burman to their folds no less than 30 times in the next 40 years. Perhaps the largest number of visits by a single Indian geographer with the purpose to comprehend the Himalayas.

The institutional base which was to foster this research interest was the Department of Geography (previously known as the Department of Human Geography) at the Delhi School of Economics. Here Professor Burman spent her entire professional life as a geographer. Picking up earlier research threads of articles on the land and people of NEFA were to be Professor Burman's first publications on the Himalayas. Situated in the northeastern extremity of India, NEFA was barely known before the Indo-Chinese aggression of 1962. Her timely and detailed study attracted immediate recognition. In a personal letter to Professor Burman, the adviser to the government of Assam wrote: "....your article, may I say so, is excellent particularly so when you have not had the occasion to visit any part of the NEFA area" (1963). The opportunity for a visit came three years later when Professor Burman availed herself of the University Grant Commission travel grant for research scholars. With this she travelled from the Apatani Plateau in the Subansiri district down to the Shillong Plateau, and the town in India which receives the heaviest rainfall, Cherrapunji. While the bond with the Himalayas deepened with this 10-week visit, Professor Burman was not content to confine herself to one corner of this vast mountain chain. To spread her wings wide a systematic collection of readings and materials on the Himalayas was next on her agenda.

Stage II: collection – 1964–1971

A case exemplifying this stage of collection is the *Bibliography on the Himalayas* which Professor Burman finalized in 1967. Mimeographed by the Geography Department of the Delhi School of Economics its relevance lies in the topical classification and notes which accompany the reference proving that a thorough reading was an ongoing part of Professor Burman's collection of published material. Even a visiting professorship to Boston University in 1968–1969 did not discourage her search for material. The literature survey had made her well aware of the paucity of published material on the geography of the Himalayas. In the USA she began to tap other sources; her letters bring forth the correspondence she had with the National American Space Agency (NASA) asking for the aerial maps on the Eastern region from the files of the Gemini and Apollo imagery (1969). Notes taken in the documentation centre and library at the National Geographical Society, Washington DC, contain the insertion…., "…. the map of the Himalayas was looked [for] – not found" (1969). Thus if the decade of the 1950s was spent in the search for an area to study, much of the 1960s was laboured to the collection and study of literature on the Himalayas. While for some geographers, secondary literature enmassed is material for writing, for Professor Burman it was just background against which the yarn for a tapestry had still to be designed and woven. Thus began the stage of analysis and conceptualization.

Stage III: analysis and conceptualization – 1972–1982

By the turn of the 1970s when the siren of an ecological degradation undermining human survival on earth was being resounded by the United Nations Conference at Stockholm, Professor Burman was busy working on a blueprint for the restoration of the Himalayas. She wanted the design for the restoration to be one based on an analytical diagnosis of the problem. What the characteristics of the Himalayan ecosystem and its resources are; how human presence impinges upon this mountain system; what the consequences of this ecological degradation are; and lastly, in which geographical unit should the Himalayas be studied and ultimately managed were questions she was keen to explore.

A multidimensional and three-pronged strategy was adopted to achieve an innovate comprehension of the above aspects of the Himalayas. These were: teaching, student research and self-research.

In 1972, when the Department of Geography set the task of reformulating a new syllabus, its teaching strength was six in number. An equitable division of the 16 courses of study in the postgraduate programme meant that Professor Burman received four in her kitty. An important distinction is that for the first time in her academic career she had the opportunity to create her independent course of teaching. Her design for the study of the Himalayas is apparent in all four courses she personally structured: (1) Ecology of the Physical Landscape, (2) Natural Resources, (3) Man and Environment, and (4) Himalayan Area Study.

The first three were foundations which strengthened the conceptual base which could later be applied to any area under study, for the unhesitant to acknowledge the role class lectures play in promoting understanding of the subject – the classroom is a soundboard for the discussion of ideas and has its place in the scheme of study. She believed any research without teaching was fraught with the danger of being unreflective. But then she also advocated that teaching without research could become a mechanical and sterile exercise. Hence an ongoing process of research on the Himalayas was very much a part of her plan.

By the mid-1970s Professor Burman began to express that only students interested in the study of the Himalayas should enrol for research under her guidance. A glance at the captions of the research work of all her MPhil and PhD students illustrates this commitment. Choosing different locales in the Himalayas, all her students pursued a central theme: the human impact on the ecosystem of the Himalayas. All of her research students were required to work on a geographical area: basin, valley, urban area or watershed.

While classroom teaching and guiding research were important ingredients, what cast the die was Professor Burman's own research on the Himalayas. Under the auspices of three research projects, sponsored by the Indian Council of World Affairs (ICWA), the University Grants Commission (UGC), and the Committee on Studies for Cooperation in Development in South Asia (CSCD), Professor Burman put together much data and information on the Himalayas. It is in the methodological organization of the material that one can see a design at work. Professor Burman was convinced that the political boundaries of the Himalayas into states or districts did not reflect its geographical reality. Almost all the published material and data in the 1960s and 1970s – until now – dealt with the Himalayas within the confines of its political boundaries. Grasping an effective geographical unity from an operative administrative unit for which data is available was thus a task before Professor Burman. To arrive at this she superimposed on the 46 districts of the Himalayas the fault lines separating the Sivaliks and the Middle Himalayas and the central axis dividing the Middle Himalayas from the Great Himalayas. These two fault lines thus provided the boundaries for the threefold ranges of the Himalayas, piercing through many districts and thus fragmenting the myth of their geographical unity. Therefore, abandoning the boundary of the 46 districts Professor Burman mooted to collect data at the tehsil (administrative area) level. In the Indian Himalayas there were 174 tehsils in 1995. Following as far as possible the dictates of the fault lines these 174 tehsils were classified into the Sivaliks, the Middle and the High Himalayas. Having understood that the Western Himalayas differed vastly from the Eastern Himalayas these zones were also classified into three sectors: Western, Central and Eastern.

Analysing data and information along these zones and sectors added novelty to the design. The task was difficult and different and no geographer has followed it – few would dare. Some salient findings of the analysis of data were drafted by Professor Burman in a report on the Development of Himalayan Resources prepared for the Indian Council of World Affairs. For many a geographer the completion of an exhaustive project and its report warrants the end of research and is usually followed by its publication. Many could argue that this is exactly what Professor Burman should have done. By now a professor and head of the Geography Department at the Delhi School of Economics, with a decade of teaching and research on the Himalayas the time was indeed ripe for her to write her treatise. But Professor Burman had other plans. The stage of analysis and conceptualization had only reaffirmed the incomplete and precarious character of the task in her mind. She moved towards the fourth stage in her design for the study of the Himalayas.

Stage IV: investigation and discussion – 1983–1992

Professor Burman had visited various areas in the Himalayas through the 1970s such as Shimla (1970), Kashmir and Ladakh (1972), Sikkim (1974), Shimla (1976), Kullu-Manali (1977), Nepal (1978) and Garhwal (1980). But the most systematic and investigative geographical fieldwork of her life was phased out from May 1983 to April 1985. In this two-year period she covered the length and width of the Himalayas. Traversing different altitudinal zones and landforms Professor Burman visited 44 districts out of a total of 46 districts of the Indian Himalayas. In these, detailed interviews and discussions were held with 250 officials. In-depth studies of 24 villages were also completed.

The terrain of the Himalayas and its climate vagaries could sap the strength of many a younger geographer but not this professor of 62 years. While a changing retinue of students accompanied her on different legs of the journey, Professor Burman showed bewildering tenacity to observe, question and note the ecological degradation of the Himalayas. That she enjoyed being amidst the Himalayas as much as she was saddened to see its ongoing environmental degradation could only be experienced when accompanying her in the field. Her alert comments on the hanging valleys, debris cones, tongues from the road, chocked streams, sheepwalk slopes brought forth the magnitude, intensity and scale of ecological problems that modalities of human development were creating in the Himalayas. Transcending the geographical scale from the macro to the micro, an integral part of the penultimate stage of Professor Burman's design was an in-depth study of a Himalayan watershed. The Kali River Watershed covering an area of 10,870 km^2 became the focus of her unblinkered attention from 1987 to 1990. Within the broad tapestry of the study of the Himalayas an intricate inlay was carved with that of the Kali watershed.[2] Thus from the zones and sectors, to the villages, Professor Burman had come to study a watershed.

Drawing from such a rich experience a hallmark of this phase was the recognition she began to attract. Although a bit late as an acknowledgement of her expertise on the Himalayas, it was a recognition of acquired knowledge and wisdom and not mere social contacts. Invitations from a number of authors and committees began to arrive. Professor Burman, along with Professors Moonis Raza and Dakshni, became a member of the subcommittee of experts to scientifically examine the issues of "Hill area delineation" framed by the Government of India (1984). From the turn of the 1980s to the mid-1990s, Professor Burman was invited to no less than three dozen seminars about the environmental issues on the Himalayas. Of these she attended only one-third. Noteworthy among these was the national seminar on *Environmental Impacts: Issues, Challenges and Assessment* at the Geography Department, Banaras Hindu University (1987); the *Workshop on Science and Technology* by the Department of Astronomy and Space, Patiala (1988), and the *Symposium on Management Issues and Operational Planning for India's*

Borders (1990). Her reputation extended to neighbouring countries and she was invited to give the keynote address at a UNESCO sponsored seminar on *Man and Environment* held at Decca University, Bangladesh (1985). At Kathmandu, she was a key resource person in the workshop on the Himalayas as part of the studies in the Cooperation in Development in South Asia. In each of these and a multitude of other seminars, Professor Burman presented her research papers.

The seminars thus became the podium for discussion. The contents of these research papers reflect the maturity Professor Burman had gained of the ecological degradation of the Himalayas. Of particular value are the following papers: *Resource Use in the Himalayas and its Ecological Implication* (1985), *Biotic Degradation in the Himalayas* (1987), *Modalities of Economic Development in the Himalayas and Their Ecological Impact* (1988), *Watershed as a Unit for Environmental Management* (1989), *Himalayan Environment and its Fragile Ecosystem* (1990), *Ecological Degradation in the Himalayan Resources Resulting from Development Strategies in Agriculture* (1991), and *Resource Management and its Ecological Interlinkages in the Kali River Watershed* (1992).

These papers reflect that Professor Burman had researched most of the questions she had set out to ask about the Himalayas. While some of these papers are published and others await publication, what is more important is that they are indicative of her finalization efforts towards the treatise. In fact, by the end of 1990s, two parallel manuscripts were under preparation – one on the Himalayan degradation and the second an integrated study of the Kali River Watershed. While sections of the first lie in various stages of completion, the manuscript of Kali has met the approval of the publishers.

Unfolding Professor Burman's design for the study of the Himalayas from the reconnaissance, collection, analysis, conceptualization, field investigation and discussion to writing confirms one fact – it would take its time. It is innovative and follows a scientific rigour. This augurs well for the nature of research required to uplift quality and scholarship of Indian geographical works. Perhaps time could have been better organized but who could foretell that the cruel hands of *yamraj* (death) would snatch a scholar's pen just when the ink is flowing full and rapidly. But here it may also be added that if Professor Burman had lived another lifetime she would still have had a cornucopia of unfinished words – so insatiable was her thirst for learning.

Thirst for learning

Born in 1921 at Lyallpur, a town 90 km to the south-west of Lahore, as a young child, Professor Burman's thirst for learning put her way ahead of the students in her class. She was 15 when she completed her matriculation and by 19 she had a graduate degree. While her father Mr Nain Sukh Gauba was a qualified engineer and barrister with both degrees from England – Professor Burman's several childhood narratives confirm that it was her simple, though illiterate, mother who goaded her into the importance of learning. Economic circumstances at home circumvented a smooth academic path. Choosing botany in school with all aims to become a doctor, the lack of finances forced her to abandon medicine and choose teaching as a profession. With a bachelor's and teaching qualification at the young age of 20, Professor Burman became a lecturer at the Khalsa College in Lahore, followed next by a year at Indraprastha Girls High School. While holding two teaching posts simultaneously – one at the Fateh Chand College for Women and the other at the Mahila College for Women – in 1944, Professor Burman's keenness to learn pushed her towards post-graduation.

Geography in pre-independent India was still struggling to gain a foothold at the university level. While Aligarh had a geography department in 1931, Punjab, the region she belonged to, only had one in 1949. Geography was new and, to Professor Burman, it was a challenge. Thus with a rather rare combination of botany and geography, she proceeded to do her masters' degree as a private candidate. In the years before partition, India was undergoing a nationalist revival caused by the freedom struggle movement. Among the youth there was high regard for courage and creativity as well as a respect for authority and tradition. For Professor Burman who enrolled as a youth congress member in Lahore in 1942 – visits to Lala Lajpat Rai's residence and her attendance at a rally for Mahatma Gandhi stayed as memories of awe till the very end of her life. While the political fervour of the times sowed seeds of nationalism, the passion for learning capped all reason and Professor Burman set sail on a converted troopship, the *Marine Adder*, for the USA in April 1947.

Finances were tight barely six months married and as a bride to be she had openly expressed her wish for any wedding presents to be in the form of cash. It is the savings from the latter which paid her and her husband Shelley Burman's fare to the United States. Recalling the trip she once expressed, "…it was a present I'd always be grateful for." Considering the 1940s, if Professor Burman dared to raise a demanding request she also worked to excel. Early in life she established the importance of both rights and duty. All through school and college she had been a meritorious student. It is her performance and teaching experience which rewarded her with a place at Clark University.

"Clark University was a powerful force in American geography during the Atwood years of the 1920s, 1930s and the 1940s", writes Marle C. Prunty in an article on "Clark in the Early 1940s."[3] In the spring of 1920 Dr Wallace A. Atwood was offered the Presidency of Clark University and the same year the Department of Geography was established. When Professor Burman joined Clark, Atwood was about to retire and Dr Samuel Van Valkenberg became the Director of the Graduate School of Geography. At Clark, Professor Burman proceeded with her PhD studies under Professor W. Elmer Ekblaw. The latter was at one time a favourite student of the environmentalist Ellen Churchill Semple. Professor Ekblaw, who was also the editor of *Economic Geography*, had an unusual training in botany, geology and geography. This made him an exceptionally valuable observer and interpreter of the natural environment. Professor Burman's botany and geography background helped her to secure a place in Ekblaw's class. Eager to learn she was quick to gain praise and thus wrote Dr Van Valkenberg, "...Mrs. Burman is one of the most brilliant foreign students we have ever had at our graduate school of geography, with one exception she had made a complete 'A' record which does not happen very often" (1948). From a graduate student with fellow rating, Professor Burman became a fellow for the academic year 1947–1948. This upgrading carried a remission of tuition fees. She also won a scholarship from the Watumall Foundation, USA. The years ahead were fruitful and tough; fruitful because Professor Burman proceeded to successfully complete her doctorate on "Geographic Problems of Land Use in Nova Scotia." Trained under masters like Professor Ekblaw, Professor Van Valkenberg, Dr R.J. Lougee, and Guy H. Burnham, her stay in the USA deepened her love and understanding of geography. But they were also hard times. Her supervisor Professor Ekblaw passed away in 1949 while she was in the process of completing her thesis, shortly after her father died in India in 1951. The dwindling economic situation added to her woes. In a letter to Van Valkenberg, who agreed to supervise her thesis after Ekblaw she wrote, "...I have already had to sell some of my dresses (sarees) to bear the cost of this thesis...due to adverse circumstances in India, we have not received any money from home since last summer" (1950).

The USA in the early 1950s offered much opportunity for research and jobs, and compared with the present, immigration was easy. In her stay of five years Professor Burman had developed a deep respect for the Americans and their style of work, but the feelings of nationalism from the days of pre-independent India had not died. Crossing the Atlantic on the Queen Mary and travelling through the European continent via the Suez Canal Professor Burman returned to India in 1952. Her contribution to India was to be in the sphere of education. The choice of the field again reflects her thirst for learning. Leaving the shores of the USA marked the end of her formal apprenticeship as a professional geographer. But as the design for the Himalayas had displayed learning was to be an ongoing process.

It would belittle Professor Burman if one thought that her domain of learning was confined only to geography and the Himalayas. The bonsai on her writing table, the flowers on her balcony, the crochet lace cover on her stitching machine and the jars of exotic cookies and pickles – which graced her simple and well-kept home – were all her talented creations. Notes on learning landscape painting, the sitar, office secretariat work, French, German and the Gurmukhi languages showed her innate curiosity to acquire skills. Files on naturopathy, colour therapy, palmistry and spiritualism tell us that her mind was also keen to capture a wide horizon. These she said were some of her hobbies. Besides the department and home, another arena which Professor Burman considered as very educational was her association with the Experiment in International Living. With Justice P.N. Bhagwati as chairman of the Advisory Board, the Experiment is a non-profitable organization involved with the orientation of foreign students to India. Professor Burman's 30-year association with the Indian Experiment began when she led a group of college teachers to the USA in 1966. After her return she became an active voluntary member of the Experiment. She was a great help in formulating and conducting orientation courses, training courses and workshops of various kinds. Her bi-annual talk on *India – Land and People* to foreign groups made her popular with a large global family. Professor Burman served the Indian Experiment in various capacities as the Community Representative at Delhi, and a member of the National Council, Steering Committee and Board of Trustees. It is her devoted work with the Experiment which gave her an office in the Vishwa Yuva Kendra after her retirement from the Delhi School of Economics.

Her multifaceted interests have been chided by many critics as one of the reasons for consuming her time. For this she argued, "the fragrance of life comes from experiencing its totality." In trying to grasp so elusive a reality called totality many unfinished words were bound to remain. It was her firm belief that reality could be gleaned by being in the environment. Here emerges Professor Burman's unstinting support to fieldwork.

Fieldwork and geography come alive

Professor Burman's description of geographical locations features or landscape had the ring of first-hand observation. In an age when arm-chair geographers have couched into the class or library, Professor Burman laid great stress on fieldwork. She advocated that fieldwork and field research is a foundation of geography and integral to the true understanding and appreciation of it. It is on her insistence that a local and long field report was made a compulsory part of the undergraduate and postgraduate syllabus of the Delhi University. Even when this compulsion was scrapped at the master's level, she still took for the Delhi Land System trip, a survey she became famous for. Her identification of the Khadar, Bangar, the meander scars, the natural levees, the secondary slopes of the Aravalis and the tors near the Jawaharlal Nehru ridge all made the physical ecology of even an urban area come alive.

Map reading in the field with the landscape before was a stimulating exercise which I experienced when in the Kashmir Valley with her. Neat and clear sketches with simplicity and truth as the main aim are inserts of many notes Professor Burman has diarized in her fieldwork. Inheriting a love for photography from her father, Professor Burman's nearly 3,000 slides and albums filled with photography tell us how keen she was to add geographical knowledge to her treasures. Aptly summing the role of fieldwork she said: "In looking one is learning, by sketching one is interpreting the landscape and by slides and photographs one can refer to texts and seek further elaboration and clarification." In all the research she undertook – may it be on the NEFA, Himalayas, Libaspur, Delhi Land System, the Damodar Valley – fieldwork was an essential component.

While Professor Burman's love for travel was a childhood passion, the professional training at Clark School nurtured and created her eye for the landscape. In her days at the USA, there was a widely based tradition of studying geography out of doors and discussions concerning geographical ideas and procedures were more vigorous in the field than within the walls of the seminar room. In fact, immediately after registration of the graduate students came the field camp. Traditionally, a location was selected within a range of 50–80 kilometres of Worcester. From this camp the graduate students went out each day to map landforms and land use.

In her later years at the Delhi University, to stall the waning interest of field work among students, she even instituted a small travel grant scholarship under the auspices of the Association of Geographical Studies.

The trio combination of an innovative design of work, a thirst for learning and importance of fieldwork were yardsticks of scholarship self-imposed by Professor Burman. This put Professor Burman ahead in many spheres of academic life. She was among the first students to opt for a botany-geography combination of study at the undergraduate level. When she began her first teaching assignment at the Khalsa College in Lahore (1941) she was younger than some of her students. In the 1940s, a decade when fewer girls pursued higher studies Professor Burman set sail to the United States for furthering her academic ambitions. When the Himalayas were little studied, Professor Burman chose to begin work on the NEFA, an even more challenging corner of this land system. In the early 1970s when the echoes of environmental problems were barely heard in India Professor Burman was teaching a full-fledged course on it. Professor Burman was also a pioneer to steer wholesome research away from urban-regional planning studies towards the studies of environmental problems, especially on the Himalayas in the Department of Geography at the University of Delhi. The forty per cent of all MPhil and PhDs submitted from 1970 to 1988 which have worked on this theme – are a legacy of Professor Burman. While she was ahead in many a scholastic thinking, a series of academic events played their contributory role in dampening her literary publications.

A series of academic events

Before Professor Burman could settle at one academic centre a series of events took her to ten different institutes for jobs in three different countries. Most of the dislocation in the pre-1945 years was due to a juxtaposition of job availability paralleled with her pursuit of studies. In this respect most of the movement was confined within the city of Lahore. Returning from the USA, the decade of the 1950s was unsettling from the standpoint of Professor Burman's career. In India it is the husband's job which decides the women's place of work, even though the latter may be far more qualified. While Shelley Burman searched for a suitable job – Professor Burman hopped with him from Delhi to Rangoon, Pondicherry, Bombay to Nangal. Geography in India in the 1950s was a discipline which only two dozen universities offered. Pondicherry and other towns such as Nangal had no geography department. The restricted distribution of geography departments was one cog in the wheel which delayed Professor Burman's entry into mainstream academic life in India.

Exhausted from bohemian life, when the Burman's visited Delhi in early 1959 Professor Burman decided to drop anchor in this city. As chance would have it, opportunity would knock; in July of the same year a geography department was started at Delhi University. Serving for a short period of six months at Camp College – then under the Punjab University, at present Dayal Singh College – Professor Burman was selected as a lecturer in the Department of Geography in October 1959, a few days short of her 38th birthday. It is in the Department of Geography at the Delhi School of Economics that Dr Burman stayed on from 1959 until her retirement in 1986. During this long 27-year tenure, while she served the department and the discipline in varying capacities a vortex of academic events thwarted her career progress. An analysis of the Table 2.1 confirms that compared to her colleague Professor R. Ramachandran she climbed the ladders of the academic rung at a sluggish rate.

The upward mobility from a lecturer to a reader to a professor can be brushed aside by many as being a superficial and outward measure of scholarship. While true, it cannot be denied that a prolonged pause between the stages can lead to dampened incentives and dissipated energies which can be a bottleneck to professional achievements.

Whereas a scholar is judged by his perception of his discipline, it is the academic environment of the department within which he functions that plays a vital role in influencing the rate at which he scales the ladder of career success.

More than half of Professor Burman's professional life in the Department of Geography at Delhi University was a struggle against a tide of academic events. Stretching from 1959 to 1973 these were years of marginalization. It is difficult to nail the exact cause which sucks one into a vortex of events

Table 2.1 Delay in academic positions for Professor S.G.Burman

Position	S.G. Burman	R. Ramachandran
Year of birth	1921	1936
Lecturer	1959	1958
Number of years	(17)	(12)
Reader	1976	1970
Number of years	(3)	(7)
Professor	1979	1977
Number of years	(1)	(19)
Retirement	1980	1996
Extension	1985	Not interested

and a close post-mortem leaves more doubts than confirmation. Yet one fact appears clear that the 14 years of marginalization was not the result of a uniform set of factors. In the first phase (1959–1965), it is the nature of the geographers which imposed limitations, while in the second phase (1966–1972) it is the nature of geography which was pursued that seems to have kept Professor Burman at bay. Both were a result of the composition of the academic staff of the geography department.

In the first phase, the department consisted of the following faculty: Dr S.S. Bhatia, Mrs Rukmini Srinivas, Dr Amrit Lal, Dr Ranjit Tirtha, Professor George Kurian and Dr Burman. The remains of some correspondence between Dr Burman and Professor George Kurian shed light on the fact that they were on different wavelengths and speaking, so to say, a different language. Personality differences created temperamental clashes. Although there were few students, the geography faculty seems to have had little unity regarding a coherent syllabus structure or a research agenda. In this setup Professor Burman had little room to move and even the performance of the others languished. The two headships of Dr S.S. Bhatia (1959–1962) and Professor George Kurian (1962–1965) fell within this phase and neither were able to accomplish the task of building a strong and viable department. Soon, except for Dr Burman, all left to seek new and more lucrative pastures abroad.[4]

While the rough winds subsided, the geography department in 1965–1966 was more like a ship without a rudder. A rather mellowed Dr Burman was unable to take charge of this vacuum. Taking charge as head in 1966 Professor Prakash Rao rescued the department only to veer it along his chosen path. The quantification, regionalization and urban studies were on the path which preoccupied Professor Rao's full attention. Known for being a team worker Professor Rao was quick to enrol Dr Ramachandran into his vision. The latter had joined the department in 1969. Professor Burman had neither the training in the field of quantification nor the inclination to pursue urban and regional planning. Further, to be part of a bandwagon knowing she did not belong in it was never her style. Thus, for the second time in her professional career Professor Burman was stranded away from the winds which would sweep her towards career success.

In this vortex of events in contrast was Professor R. Ramachandran. Experienced in the field of quantification from his days at the Indian Statistical Institute and having trained in urban geography under Professor Raymond Murphy at the Clark School were timely assets for Professor Ramachandran. Further, he was undoubtedly Professor Rao's black-eyed favourite and thus quickly rode the tide of success from lectureship to readership and later, after Professor Rao left, to professorship. Professor Prakash Rao's tenure in office provided stability and continuity to the geography department, but his intellectual feudalism so dominated the department that for seven years all PhDs, all research projects and much of the postgraduate syllabus revolved around the theme of quantitative revolution and regionalization. At the time when Professor Prakash Rao left the department (1972),

Dr Burman was nearing her 50th birthday. She was still a lecturer while a 15-year younger Dr Ramachandran was in the offing of becoming a professor. However, despite these differences what becomes remarkably noteworthy is the camaraderie both displayed while steering the department through difficult times.

When prefixes like professor are dissolved as being unimportant there emerges the true personality of a scholar. Both Professors Burman and Ramachandran put this belief into practice. What is more when the younger staff Dr C.P. Singh (1972), Dr S.K. Pal (1966), Dr Noor Mohammad (1976), and Dr B. Thakur (1980) joined the faculty they too were absorbed into the family-like spirit of working in the department.

Professor Burman got more than just a breathing space in the independent and self-committed functioning of the department under the leadership of Professor Ramachandran (1973–1979). In 1979, she gained her professorship and became Head of the Department of Geography. All of Professor Burman's research projects, research students, seminar participation and favourite courses to teach belong in this phase. Perhaps if the winds of change had rushed in a little earlier, more of Professor Burman's publications would have reached the shore.

It is difficult to unravel if events shape our destiny or destiny shapes our events. A firm believer of the *karmic* cycle Professor Burman, amidst a vortex of events, continued to perform many untold works.

Untold works

In Indian universities there are no rosters that academics have to follow. The amount and kind of work an academician does is therefore entirely a function of his life perception. When Professor Burman joined as a lecturer, the Department of Geography was only four months old. The building of the Delhi School of Economics which had been completed in 1956 was overpowering and spacious. But in 1959 when geography was introduced under the umbrella of social sciences there was no place for a separate department in the school. A makeshift cartographic lab-cum-lecture room was set up in the present Delhi School office. While the head of the geography department occupied a room next to the director's office, Professor Burman shared a room with Mrs Rukmini Srinivas down the corridor. If space and financial resources were at a premium, even the student number threshold needed to sustain a postgraduate programme was lacking. The unpopularity of geography at secondary level in schools was seen to be the reason for this state of affairs and this was further thought to be the result of a lack of geography teachers in school. To cover this lacuna an evening diploma course in Postgraduate Teaching of Geography was introduced. Though the diploma course lasted for only a few years it was responsible for eliciting interest in geography at school level. It is the grassroots hold of geography in the schools which later grew to fuel the increasing number of students filling the undergraduate and postgraduate geography classes.

Professor Burman was the key person entrusted with this programme and many of her earlier timetables record the evening classes: 4–6 PM.

When the geography department was shifted out from the main building of the Delhi School to its present locale in the teaching block in the early 1960s, it had little infrastructure to support a research and teaching programme. Purchasing cartographic equipment, compiling a department library, planning a seminar room are works Professor Burman performed as a pioneer in the department. Acknowledging that the geography department drew many advantages because of its location within the Delhi School premises Professor Burman extended her efforts to the maintenance of the entire school.

That Professor Burman plunged headlong into contributing in this direction can be seen by the number of committees she chose to be a member of. Her enrolment in the Library Committee, Canteen Location Committee, Master Plan Committee, Cafeteria Maintenance Committee, International Students' Service Cell, and Garden Committee all display the interest she took in varied aspects of the care of the school. This extended her interaction with faculty members outside the geography department and won her many friends. Neat files of the circulars and follow-ups of these varied committees tell us that work recording the minutes of these committees at times took many days. As an example, in 1967 as a member of the Library Committee, Professor Burman chalked a detailed plan of the geography department's needs for the library. A prioritized list of books and journals which are essential for a research and teaching department were valuable additions made by Professor Burman's efforts. It is from these embryonic beginnings that the Ratan Tata Library can today boast of 8,000 to 10,000 geography books and 96 international and national geographical journals today. Similarly the many avenue trees, shade trees, and flowers which add seasonal colour to the school is the continuous work of the Garden Committee of which Professor Burman was a keen participant.

Such works are no small feat and go to build the work ambience of an institution. Posterity will pay the price if these works are neglected and remain untold in the name of a waste of intellectual time, or the excuse that brilliant intellectuals are not always effective builders or even managers of institutions is offered.[5] What perhaps perpetuates the above thinking is the unrecognized place such caring works merit on our list of accomplishments. If publications remain the only acid test of scholarship then one is compelled to ask – how long can an institution survive with growing clumps of parasites? Professor Burman was the archetypal moralist, the epitome of those who care for an institution which is not theirs and offers them little later. Such scholars are rare today.

While remnants of physical artefacts are clues to the untold works of predecessors, what remains oblivious is the work one performs in upholding academic standards in a discipline. Concrete evidence of this work is difficult to gather and has to be experienced more than expressed. Here I will take the liberty to share my own.

Professor Burman was at the peak of her academic career when I joined the Department of Geography as a postgraduate student during 1979–1980. She had recently attained professorship and was to take over headship of the department from Professor Ramachandran the following year. Compared with the present the department was small. There were six on the faculty with around 28 students in each year. The MPhil programme had recently been introduced in 1977 and this group added conversations in the corridors about research.

Professor Burman offered three courses for teaching: *Ecology of the Physical Landscape, Man and Environment* and *The Himalayas.* My first interlude with her in the lecture on Ecology of the Physical Landscape was a rocky one. While she stressed the most basic and simple concepts needed to comprehend *Ecology of the Physical Landscape,* I was keen to hear models and theories, strewn with names of geographers and journals. While I wanted to fly – with little training – her entire class lectures were oriented to plant our feet in the grounding concepts. The result was that I failed the course. It was only the second time while repeating the course when I was willing to be in a listening space that I began to see how artfully she wove the interrelationship between so many a geographical variable to create our understanding of the landscape and the Himalayas.

Her painstaking comments at times weeded out my most favourite and what I had thought my best portions in the tutorial. Her thought provoking questions kept me busy in the Ratan Tata Library which in those days remained open until 10.00 PM. She stressed the outline method to define the bold structure of our assignment. I was later proud to know that among her students she thought I had quite got it. Reticent to shower praise, all I can say is her training stood me in good stead. From an abysmal B grade in my first year of the postgraduate programme I not only achieved an overall A grade in my post-graduation but I had developed an immense interest in the theme of ecology and the environment. Though we both shared a love for the same theme, both of us had reservations about getting along in so crucial a role as student-supervisor. While I think it was courageous of me to opt for Professor Burman as the only choice for a supervisor, in retrospect it reflected her scholarly attitude to take me on: obviously, merit had won over personal likes and dislikes. Thereafter in our 15-year relationship she was to be a most affectionate guide and upholding standards was to be her overriding motto.

To uphold standards detachment is obviously a valuable asset which Professor Burman possessed even though it did not always work in my favour. When the post for a lecturer was advertised in the Kirori Mal College in 1985 Professor Burman was the expert from the geography department. I had recently completed my PhD under her supervision. The other contester for the post was Dr B. Marh. He was a PhD from the United States, with a book, research papers and was holding the post of a pool scientist with the ICSSR. Professor Burman, along with other members of the selection committee, settled for Dr B. Marh. I will never forget her comment afterwards: "This is a Selection Committee and to uphold standards only merit should count." Rather hurt, I mumbled in all modesty, "but I am not bad." Professor Burman's reply, though filled with empathy, was "…he was better."

Her very presence lent an air of upholding standards and she was willing to hold fort until the very end. She sat through an endless number of selection committees. The list ranges from selection committees for jobs as cartographers, lecturers and schoolteachers. For scholarships, travel grants and research projects, she was a member of the selection committee for the UGC, ICSSR, CSIR and the NCERT. She was a member of the Examination Board for many geography departments in India and also set papers for the Rajasthan State Commission and the UPSC. Working for these committees we all know is a lot of labour for a paltry sum. Yet Professor Burman willingly took on these mostly untold works and academic chores. As Professor Burman said, like Alice in Wonderland, these selection committees are the most realistic peephole to comprehend where geography and geographers are heading in India. But by the end of her academic life, unlike Alice, she felt more in a blunderland; a consequence she said of the overall degradation of the environment. But she was no pessimist. She had faith that human will is a force of unseen proportions which needs to be tapped. Education, she felt, was the cornerstone. Thus, she lived her life until the age of 74 devoted to the cause despite unfinished words and untold works.

Webster's Third New International Dictionary (1961) defines the word 'scholar' as follows: "a learned person (especially) one who has the attitude (as curiosity, perseverance, initiative, originality, integrity) considered essential for learning"… unrelated to individual profit. It does not even mention publications, let alone debate over numbers. Professor Burman, my guide, was a scholar par excellence.

Acknowledgements

I would like to express my thanks to R. Ramachandran, C.P. Singh, and B. Thakur in the Department of Geography and Shelly Burman, Professor S.G. Burman's husband, for valuable information in the completion of this research.

Notes

1 Mr Tarlok Singh served as Deputy Secretary of the Indian Planning Commission at that time and later worked as its Member.
2 The Kali River Watershed lies in the Central Himalayan sector. The river known as the Mahakali in Nepal forms the international boundary between India and Nepal. Professor Burman conducted research in the administrative portion of India which covered the entire district of Pithoragarh, the Bageshwar Tehsil of Almora and a very small section of the Almora Tehsil.
3 Marle C. Prunty, "Clark in the Early 1940s," *Annals Association for American Geographers* 69 (1979): 42–45.
4 R. Ramachandran, "Department of Geography – Three Decades of Growth," in *XI Indian Geographical Congress Souvenir*, National Association of Geographer, India, Department of Geography, Delhi School of Economics, October 1989.
5 Andre Beteille, "My Formative Years in the Delhi School of Economics (1959–72)," in *D. School: Reflections on the Delhi School of Economics*, eds. Dharma Kumar and D. Mookherjee (New Delhi: OUP, 1995), pp. 53–67.

3 Opportunity a bit late...
Death a bit soon

C.P. Singh

Professor Chandra Pal Singh passed away on 5 December 2000. It was a Tuesday afternoon and he was in the throes of his work at the Department of Geography, Delhi School for Economics where he had served as a geographer for 28 years. In just six months he would have retired anyway and perhaps left the portals of this institution. For the kith and kin of the person concerned, the eventuality of death even at age 100 would seem soon, but Chandra Pal's pace of work and his apparent robust health persuaded even the most adamant critics to voice in dismay that he had died a bit soon. While assessing the life of a professional, it is often not age but pending work that qualifies 'a bit soon.' Death liberates a person from judging and questioning either its time of arrival or the remaining tasks. Why should one pen a biographical analysis? Life is about learning, and much can be accrued from a study of the lives and works of others. Further, it helps in drawing objective conclusions about the contributions of a scholar. It is framed within the voyage of life that the context of work emerges. Most importantly, a careful study of the lives of others provides landmarks for self-referral. They are effective mirrors for one's self.

Journeying a life and mapping its nuances requires a rich repository of carefully sifted data. Archival and official records, personal diaries and letters, notes and publications, and searching perspectives and perceptions, form an essential base. While accurate plotting and appropriate labelling are virtues, at its ultimate, the power of a map depends upon its interpretation and there can be numerous versions for the same data. Each to his own; here is mine!

What is the landscape of Chandra Pal's life? It is not a landscape of a scholar towering with awards, scholarships and positions. It is also not a landscape of dense publication of books, papers and monographs. Neither is it one of a fertile legacy of students carrying forth ideas and thinking. It is not even a scarred and degraded landscape of financial frauds, politicking, repetitive publishing to add to quantity, and awarding research degrees to the unfit candidates. It is a polycyclic landscape. From foreign area expertise, land use specialization to electoral geography to political geography of India are all changing features of Chandra Pal's career graph. It is a landscape that attempts to build a scenario of research expertise and then abandons the entire exercise. It is a mixed landscape where, in some places, the terrain is monotonous and others well built. His most coveted acts are positioned in the penultimate part of his life, many works are incomplete and, hence the conclusion, Chandra Pal died a bit soon. There is, however, a flip side; few realize that opportunity came to him a bit late and therefore death seemed a bit soon.

Opportunity a bit late

Chandra Pal was appointed as a lecturer in the Department of Geography, Delhi School of Economics in 1972. He became a reader – equivalent to an associate professor in university academic ranks in the US – in 1983, a professor in 1993 and chairman of the department in 1999. So, he was a lecturer at age 31, promoted to a reader at 44, professor at 54 and chairman at 60 years. The data of many university teachers could be collected to establish the average years of reaching these levels. Here, a simple comparison with another colleague, who has since retired from the department, is used. Chandra Pal was only three years older than R. Ramachandran. The curriculum vitae of R. Ramachandran brings forth the fact that Chandra Pal's promotion to these posts lagged by a decade (Table 3.1). These steps of the academic ladder could be disregarded as trivial but no one can deny that many opportunities appertain to rank. Even if he had lived longer, Chandra Pal would have been chairman of the department for a brief period of only two years, while R. Ramachandran ruled as chairman for 15 years. Chandra Pal could avail himself of only one sabbatical, due to delayed professorship, while Ramachandran enjoyed two study leaves. Benefits do flow from positions. Comparing lives, many say, is inappropriate but a qualifying statement does draw strength from a correlative analysis.

Freeing the assessment of opportunities from the tight frame of promotions it could be asserted that 28 years is a long enough duration of time to create indelible imprints on a profession. After all, on the yardstick of

Table 3.1 Lag in opportunities

Position	C.P. Singh	R. Ramachandran
Year of Birth	1939	1936
Lecturer	1972	1969
Age (Years)	33	33
Reader	1983	1970
(Age)	44	34
Professor	1993	1977
(Age)	54	41
Chairman of Department	16 months*	15 years

* It might have been two years had Chandra Pal lived.

tenure nearing the age of 62 one is expected to complete the many academic tasks of publishing research, tutoring students, handling projects, attending seminars and chairing the department along with a plethora of other activities.

One could subscribe to a different view and state that opportunities have a uniform distribution and one has to work to catch them. Life's trajectory is, however, rarely smooth and providences are not posted at frequent or appropriate intervals as per one's needs. Life experience teaches that determining what one gets is not entirely in the hands of a single individual: circumstances, awareness, perhaps karmic destiny among a host of other factors, act out their roles. It is difficult to pinpoint all the causes that delay opportunities. In Chandra Pal's case, the overwhelming factors seem to be his search for a destination, the prevailing academic canopy and attempts to carve a parallel identity.

Search for a destination

A critical prerequisite to reach a destination is to know where to go. Destinations for academicians are not outwardly positions and laurels; what is significant is identifying the thrust area of research towards which a long-term commitment is made. This could be a theme or a region; for geographers it is invariably both. The custodians of universities also place faith in a specialist. Let's agree that channelling energy in a single direction yields fruitful results. Identifying a goal soon is beneficial and selecting a relatively appropriate one is an advantage. Before anchoring political geography as his final objective, Chandra Pal's life chart shows that he discarded two themes of research, namely, foreign area ken and land use studies. The cost of this digression is a decade and a half of delay in availing opportunities.

Foreign area expertise

Chandra Pal took upon himself the task of acquiring expertise on East Africa. Visas in his passport confirm that he made eight trips to Africa in 13 years from 1962 to 1975. During this time he also resided for five years in Addis Ababa, the capital of Ethiopia. Learning the Amharic language and selecting Shewa, a province located in centre of Ethiopia, as his area of research for a doctorate gave him knowledge of the region. Bent on carving a foreign area expertise, Chandra Pal returned to India and introduced a course to teach about the region of sub-Saharan Africa. Stocking the library with books and journals on Africa were the first announcements of his commitment to the region. He continued to bond with Africa even after his doctorate and took a three-month grant-in-aid research to Nigeria in the summer of 1973 and a four-month fellowship under the auspices of the Association of Commonwealth Universities to collect data on Ethiopia and Tanzania in the summer of 1975. His association with colleagues, namely, Dr Kathy Baker a specialist on East Africa at the School of Oriental and

African Studies is proof of his inclination towards the region. Though his doctorate remained unpublished, most of his earlier publications focused on Africa. *Review of Rural Development and Bureaucracy in Tanzania: Case of Mwanza Region* (1975) *Strains and Stresses in Ethiopia* (1976), and *Review of Managing Rural Development: Ideas and Experience in East Africa* (1976), are evidence of sustained interest in the foreign area. Participating in a conference on *International Struggle Against Apartheid*, New Delhi, in 1979 was an offshoot in a similar vein. While at the University of Delhi, he established links with the African Studies department and worked in various joint committees with them – the department of African Studies is in the faculty of Social Sciences at the University of Delhi. Established in December 1954 it is an interdisciplinary area study centre. Many African students registered with the foreign student cell of University of Delhi came to seek his counsel. His first doctoral student was Chidobel Offia who hailed from Nigeria.

The reasons for Chandra Pal's keenness on Africa are not difficult to discern. Understanding the innumerable similarities of the characteristics and problems which beset countries in Africa and India, was a major motive. A teaching assignment in Ethiopia gave him a feel for the land and people. His enrolment at the School of Oriental and African Studies as a doctoral student put him under the supervision of Dr J.H.G. Lebon, an authority on Africa. The unexpected death of his supervisor resulted in Dr J.A. Allan taking over his supervision. While there is no doubt that a wider geographical understanding is aided by seeking a foreign area expertise, this is often deterred by limited funds for travel and subsistence and the need to learn an unfamiliar language, as a result of which no researcher enrolled under Chandra Pal in furthering the study of Africa. Meanwhile Chandra Pal's interest in this region also waned.

Abolishing the course from the syllabus he did not write a single line or ever visit the continent again after 1979.

Umbilical cords with a doctorate are often too strong to sever. Doctorates are grounds for nurturing research acumen. Unfortunately, for many this first love becomes a long lasting passion or they grow out of it late. Often in the works of geographers although a region of study is discarded, the theme lingers. Cloning the goals of research imbibed in the doctorate into a different region thus becomes an agenda. Such was the case with Chandra Pal.

Cloning the doctorate

The core objectives of Chandra Pal's doctorate were the utilization of land by the people in the cropland area of Shewa in Ethiopia. Back home in India, he pursued similar agriculture and land use studies. When war broke out with Pakistan in 1971, Chandra Pal was busy collecting data on crops, agricultural infrastructure and productivity in the villages of Bulandshahr, a district in Uttar Pradesh. The enthusiasm to duplicate his doctorate from Africa to India was honed in the year he spent working as a research associate with a team of geographers from South Asian Oriental Studies in a rural part of his own country – this was sponsored by the Social Science Research Council of the UK. The project director for the Bulandshahr Study 1971–1972 was Professor J.A. Allan of the School of Oriental and African Studies, University of London. A four-month teaching grant from the University Grant Commission in 1972 gave him financial support to study the changes in agriculture in the field of Delhi. Documented from the archives and fieldwork maps and notes on the land use in rural Delhi form a sizeable stack of his numerous unpublished notes. Some of his publications, for example, *Agricultural Labor in Chirchita Village, Bulandshahr and Uttar Pradesh: Labour Availability and Peak Demand* (1975) and a paper presented on "Gandhian Ideology in the Transformation of Rural Habitat in the Third World Countries" (Varanasi 1978) spearheaded similar goals. Examining the relationship between agriculture and industry also falls within the ambit of land use studies. An 18-month research from 1976 to 1978 on the sugar industry in Western Uttar Pradesh provided Chandra Pal with forays into this aspect.

Two parallel perusals, Africa and land use, a hangover of the doctorate, consumed 18 years of Chandra Pal's academic tenure. They engaged him but failed to give him a sense of purpose or identity. The result was that he dumped his interests on Africa and discarded any attempt to clone the doctorate. The prevailing academic canopy also cast a shadow.

Academic canopy

The Department of Geography had entered its teens when Chandra Pal joined as a lecturer in July 1972. Founded in July 1959, the department was exactly 13 years old. In this 10-year plus existence, the department had failed to stabilize. A large number of the earlier crop of geographers responsible for its creation had fled to seek more lucrative avenues elsewhere. In its initial phase of establishment, the department saw a turnover of no less than nine faculty members, one every 17 months. Of these, with the exception of Dr Savitri G. Burman and Saroj K. Pal, all had been transient members serving the department for short periods. (Table 3.2) Long-term association and in-born commitments are prerequisites for creating stability in any given department. Teens are times of turmoil. The department was no exception.

The constituent strength of the department was a meagre four at the end of 1972 in which Chandra Pal was a recent entrant. It was R. Ramachandran and S.G. Burman who shouldered much responsibility to ensure stability. By temperament S.K. Pal was a loner and preferred to work in isolation, but as one knows it takes all types of people to make this world. When Chandra Pal joined, he gravitated towards R. Ramachandran and S.G. Burman; the trio were a team which committed all their professional years towards nurturing the department. The years saw the joint pioneering of student classes and building resources. While the active role played

Table 3.2 Tenure of faculty members in the Department of Geography, Delhi School of Economics, 1959* to 1972**

Faculty	Tenure in years	
Rukmani Srinivas	13	
V.L.S. Prakasa Rao	7	
Amrit Lal	5	
Ranjit Tirtha	5	
Shyam C. Bhatia	4	
George Kurian	4	
Krishna K. Khattu	2	
Savitri G. Burman	13	(In the dept. until 1985)
R. Ramachandran	3	(In the dept. until 1985)
Saroj K. Pal	6	(In the dept. until 1985)

* Year of establishment of the Department of Geography.
** Year Chandra Pal joined the Department of Geography.

by R. Ramachandran and S.G. Burman in the affairs of the department have been widely recognized what is less known is the supportive role played by Chandra Pal, offering views forthrightly when sought. They too serve who stand and wait. Although they occupied a common habitat all three had a different position. R. Ramachandran came to occupy the towering layer, followed by S.G. Burman and Chandra Pal yet to make a mark. It was the intellectual tenacity and fervour of urban and regional planning and the opportunity to chair the department which explains the dominance of R. Ramachandran, while S.G. Burman's expertise on the Himalayas and her seniority were an advantage. The hallmark of a canopy is that it moderates and creates a conducive environment for all to flourish. While it may seem that the taller species enjoy most resources it should not be forgotten they also create the shade and provide much wanted nutrients to anchor the system. Very few would know of R. Ramachandran and S.G. Burman's critical roles in channelling Chandra Pal's attention towards political geography. Thus, wrote R. Ramachandran "Dr Chandra Pal Singh's initial interests were in agriculture and land use studies. A few years ago, he changed his focus of research partly on my advice to political geography and in particular on electoral geography in India." (Letter of Recommendation for Commonwealth Fellowship, 15 November 1985) Gratitude is an obvious part of the character of Chandra Pal and he acknowledges, "Dr S.G. Burman and Professor Ramachandran asked me to work on this branch of geography during one of the informal lunch sessions in 1976." (In Preface of the research project *The Seventh Parliamentary Elections in India: A Geographical Analysis*, 1980). While good advice comes the way of many Chandra Pal was among the few to heed it.

Studying political science at undergraduate level and the turbulent political environment of India were factors which cemented his interest. The emergency that was clamped on 26 June which led to the defeat of Indira Gandhi as prime minister of India, ignited his interest in such aspects of political geography.

> From a personal point of view, the radical transformation of Delhi and its newly emerged face under the short-lived emergency under Mrs Gandhi, and the subsequent events swept me off my feet and brought me into political geography. Two Indian elections two within three years, were also experiences which helped me to understand the mosaic of politics at subnational level – I am convinced that politics and political processes were the most powerful elements in shaping the spatial patterns and geographies of different parts of the earth.
> (Letter to Professor D.B. Knight, Department of Geography, Faculty of Social Sciences, Carleton University, Ottawa, Canada. The letter is not dated but I think it was written in 1987.)

These were not the last of the influences of the prevailing environment. Two posts for readers were advertised in the department in 1980. Chandra Pal was not selected. It is often said that events shape our lives – but it would be nearer the truth to say that it is our interpretation of the event which moulds our reactions. For Chandra Pal it was a moment of deep anguish.

Carving a parallel identity

Of course, environment shapes us, but we also shape our environment. Jolted and hit hard by the apparent blow to his ego caused by the loss of a readership, Chandra Pal catapulted from a state of gloom into a position of identity. He wore two hats in the 1980s by acquiring the directorship of an industry and by establishing a name as a political geographer. As director of an expanding industrial concern which manufactured quartz clocks, Chandra Pal took the business to new heights of profit. Rewarded with swift success, in two years, he attracted the directorship of an additional firm, which manufactured black and white television sets. He taught industrial geography at the department and carried his lessons of theoretical knowledge across the applied field. These were other perceptions

> Chandra Pal's keen interest in National Development can be seen from the fact that All India Electronic Clocks Manufacturing Association elected him as their first Honorary Secretary. In this position, he has been able to mould some of the industrial policies of Government of India recently in pulling down trade barriers.
> (A letter of recommendation by Professor R. Ramachandran, 1985.)

Acquiring the directorship of private industry was criticized by some, envied by few and ignored by others. While Chandra Pal did attract criticism, he was too busy simultaneously pursuing an identity as a political geographer.

As a political geographer

When the *Who's Who in Indian Geography* was published in 1981, it contained a list of 31 political geographers, among which was Chandra Pal. There are a number of other ways in which an academic identity can be created. Publishing research, attending seminars, organizing conferences, teaching the course, undertaking research projects, chairing commissions and guiding students are others. One could select a single method but Chandra Pal adopted a multi-pronged approach.

R.J. Johnston, a professor at the University of Sheffield, stated, "the world political map is the most familiar of all geographical maps, yet paradoxically has been the least researched in recent times." The study group on the world political map was constituted by the apex body of worldwide geographers, the International Geographical Union, at the meeting of its General Assembly in Paris in August 1984. R.J. Johnston became the first chairman of the study group which was raised to the status of commission when the congress met in Sydney, Australia in August 1998. The commission enrols only 11 full members from different countries of the world. Chandra Pal was among this small chosen team along with political geographers like J. Minghi of America, Wang En Yong from China and R.J. Johnston from the UK. This was a good time for political geography – *Political Geography Quarterly* was upgraded as a journal, *Political Geography*, in 1992. Frequent contact with likeminded scholars were beginning to develop. A communication from Dr Geoffrey Parker, head of the European Studies Centre at the University of Birmingham stated "I have been awarded the Aneurin Bevan Fellowship for the purpose of making a study of Indian Political Geography – and would like to meet you for discussions in my forthcoming visit to India" (1998). A request from Professor P.J. Taylor's to Chandra Pal to write an epilogue on Indian perspectives for his book *Political Geography of the 20th Century* are evidence of this.

Along with an international platform he was keen to network with geographers at the national level. To knit them together his plan was

> ... to give impetus to the study of political geography in different universities of India and further to those in South Asia as a whole, I have arranged with [the] National Association of Geographers, India to organize an annual meet of political geography along with its annual Congress – my effort will be to link this academic activity with the International Geographical Union Commission of world political map. (Letter dated 20 May 1991 to Professor D.B. Knight, Faculty of Social Sciences, Carleton University, Ottawa, Canada; Chairman of the Commission on world political map.)

A commission on politics and environment was set up when the National Association of Geographers, India met in December 1991 at Magadh University, Bodh Gaya, in Bihar. Chandra Pal was instrumental in its formation. He also organized an International Conference on Politics and Development at the Department of Geography, Delhi School of Economics, in November 1990. An exchange with various political geographers across India like R. D. Dikshit (Panjabi University), S. Mehta (Panjab University), R.N.P. Sinha (M.S. University, Baroda), Manorma Sinha (Allahabad University), and J.C. Sharma (Himachal University, Shimla), had been forged and can be seen in the correspondence on different academic issues.

Presenting and publishing papers at different forums formed an additional agenda to reinforce his identity as a political geographer. "Spatial Variations in Electoral Participation in India" (1980), "Nature and Scope of Political Geography"(1980), "Political Regions of India"(1981), "Geography and Electoral Studies"(1981), "Indian Electoral Geography"(1981), and "India and the New Global Politics" (1991) belong to these efforts. Writing for leading national newspapers and interviews on the television were also used to reach a wider audience. Articles with a number of maps on themes like, "Why they Vote the Way they Do," "How to Gauge the Voters Mood," "The Voter is getting Choosier," in the daily *Statesman* (1985) belong to this category. Effectively exchanging views with political geographers across academia were planks which fostered Chandra Pal's identity as a political geographer.

The decade of the 1980s were years of great struggle for Chandra Pal. Carving two identities of very opposing requirements was like rowing two boats in different directions. The task sapped his energy and health. Chandra Pal suffered a heart stroke in the summer of 1991. A serious illness warns that life patterns need to be changed. Academics beckoned the now 52-year-old Chandra Pal towards a single tract devotion. Devolving from industry, a more sedate and wiser Chandra Pal settled with a desire to crystallize his ideas on political geography. He had identified his academic destination. Further, the experience of life had equipped him with two additional strengths; first, the ability to scale contrasting milieus, and second the influence of the British.

Scale of contrasts

Born in the village of Bitoada in the heart of the rural countryside of the state of Uttar Pradesh on 2 July 1939, Chandra Pal's family shifted to Shamili early in his childhood. Situated in the Muzaffarnagar district, back in 1941, Shamili was a small settlement of 12,000 people. Lying within the commercial sugar cane belt of India, the three commodities which distinguished Shamili as one of the few progressive settlements were the agricultural implements, bullock carts and sugar mills.

Retired as a soldier from the British army, Chandra Pal's father took advantage of the geographical location of Shamili and not only farmed, but opened a small shop selling agricultural implements, seeds, fertilizers and animal feed. Chandra Pal's primary school education was in Shamili. For his later years of education he moved to Jain Government School at Muzaffarnagar, a town 38 km from Shamili. In 1959, he graduated in geography, political science and English literature from Balwant Rajput College at Agra, the town famous for the Taj Mahal. From the heat and dust of the Indo-Gangetic Plain, the next academic halt was the salubrious environment of the Vindhyan ranges in the heart of India. Established by the philanthropist Dr Hari Singh Gaour in 1946, it was in the Department of Geography, Sagar University, Madhya Pradesh, that he acquired his postgraduate degree in 1961. It needs to be reminded here that in the early 1960s, postgraduate level geography was taught at only a handful of universities in India and Sagar was among the leading ones. After graduating, Chandra Pal returned to Dayal Bagh College, Agra to obtain the degree of Bachelor in Teachers Training. His final city of formal education was London, at the time a city of eight million people, where he completed his doctorate in 1972 at the University of London with fieldwork based in Ethiopia An ambitious student in search of learning had crossed a large sweep from a village to a town to a city to megalopolis. Academic pursuits took him through not just a hierarchy of settlements but three different continents; Asia, Africa and Europe. Scaling contrasts demands adjustment and teaches adaptation.

Chandra Pal's mother never attended school while his father studied up to the middle school level. His elder brother studied until higher secondary, but his sister never went beyond primary school. Despite his rural, non-academic background Chandra Pal succeeded in attaining his doctorate and followed a lifelong profession in teaching.

Raised in the rural areas of Uttar Pradesh, Chandra Pal lived in the cosmopolitan environs of Delhi, yet he never lost his love for the countryside and pride in being a Jat. Though he was proud of his heritage he was not parochial. It is worth noting that while appointing faculty members or cancelling registration of students not up to his standards, many who bore the brunt belonged to his very own community. To scale pride from performance is a dying wisdom in the institution of universities in India today.

The ability to straddle contrasting milieu was to be his hallmark in the years to come. Jumping from academics to industry, and land use to political geography are the obvious cases. Being able to carve international, national and local links with political geographers also speaks of a similar scale of contrast. Such ability helped him capture a kaleidoscopic range of political processes and events. Examples of global political events that were influencing local issues were often used in his class and lectures. A collection of texts and readings on the history of India, along with loads of newspaper clippings, backed more recently by downloads from the Internet tell us that linking the past with the present came to him with ease. In the latter years, as chairman of the department, it is perhaps similar sweeps which motivated him to carry the department from the morass of archaic teaching equipment to modern information technology. The mix of guests, from farmers in the small town of Shamili, to Kathy Baker from South Oriental African Studies, to Sahib Singh Verma the Chief Minister of Delhi in the photo album of his younger son's wedding, are proof that even his friendships ranged across a wide scale and endured over a long time. He often said that he was comfortable with all in all environments – little wonder that when, in 1986, he met the British prime minister, Edward Heath and shook hands with Queen Elizabeth II, it was not an event to boast about, though the impression of the British was to endure a lifetime.

British impression

The near impeccable command over the English language, flair for extempore speeches and lectures, the polished manners, conservative habits of dress and passion for Mackinder's Heartland Theory[1] shows that Chandra Pal remained under the spell of British style and magnanimity. Despite a rustic background, reading English literature was a hobby for Chandra Pal. Hailing from a rural and government school education, to overcome the drawbacks of the language, he burnt the midnight oil to master the popular *Comprehensive Grammar of Wren and Martin*,[2] a classic which scares even convent-educated me!

It was his tutelage under Professor Muzaffar Ali which fanned his interest to reach England. Ali joined Sagar University as professor and head in 1957. Chandra Pal enrolled at Sagar as a student in 1959 to 1961. Muzaffar Ali had completed his doctorate under Professor E.G.R. Taylor at Birkbeck College, University of London in 1939. He obviously recognized in Chandra Pal a talented and enthusiastic scholar with a future and convinced him to move to London. The School of Oriental and African Studies, a part of University of London was where Chandra Pal registered in 1966. Located in Bloomsbury in the heart of London, the school is close to the British Museum and British Library, Covent Garden, Westminster and the city. These were attractions Chandra Pal used for making progress in education. When I visited the School of Oriental and African Studies in 1999, Professor J.A. Allan, Chandra Pal's supervisor, remarked, "Students of the like of Chandra Pal in dedication and excellence are becoming rare at the School". He became a Fellow of the Royal Geographical Society, London in 1969. His stay in London was a period of apprenticeship and it left a deep imprint on him. Though London was a joy, Chandra Pal's heart lay in India. Without even the promise of a job in India, he returned with his family to settle in Delhi in 1970. But his sojourns to England continued for a lifetime.

As a participant in the second Indo-British geography seminar which was held at the Centre of South Asian Studies, University of Cambridge in 1975, Chandra Pal reinforced links with R.W. Bradnock of the School of Oriental and African Studies, G.P. Chapman and B.H. Farmer, University of Cambridge and M.J. Wise from the London School of Economics. After the International Geographical Congress held in Paris in 1984 he revisited

England. Chandra Pal was nominated for the Commonwealth Foundation award by the Commonwealth Geographical Bureau and was in England for five weeks in 1986. With the broad aim of increasing interchanges between professionals of different fields the Commonwealth Foundation nominates a dozen people every year from within all the Commonwealth countries. In England he met R.J. Johnston and Professor R. Bennett of the London School of Economics and also visited the Royal Geographical Society and Department of Geography at the School of Oriental and African Studies. His ties ran deep and he remained South Asia representative for the Commonwealth Geographical Bureau from 1984 to 1992 and, under its auspices, organized a conference on Urbanization in Developing countries in Delhi in 1987. In 1993, he participated in the Indo-British Geography Seminar on Environment and Development, held at the Lal Bahadur Shastri National Academy of Administration, Mussoorie. The British also left a deep impress on Chandra Pal.

In spite of Chandra Pal's ability to adapt and comprehend contrasting realms and acquire training at the best of British universities, the opportunity for the crystallization of ideas happened rather late. Though Chandra Pal was appointed professor in 1993, it was during a year of sabbatical in 1997 which he spent in America that he settled into a much needed life of quiet reading and researching for his book. On his return he often expressed to me his urge to finish his book, "I want to devote a year or two towards the completion of my book on political geography which shall show the relationship between history and geography in shaping the political processes and patterns in India." But the winds blew in a different direction. The Executive Council of Delhi University took the decision to extend the date of retirement for university teachers from 60 to 62 years resulting in Chandra Pal taking over as chairman of department at the end of June 1999.

Chairman of department

Chandra Pal was the twelfth in the series of professors to chair the Department of Geography at the Delhi School of Economics Some chairmen carry a vision about the contribution they would like to make towards a department. V.L.S. Prakasa Rao, for example, called it the Department of Human Geography and wanted to create a centre of excellence in urban and regional studies; in 1976 R. Ramachandran deleted the adjective human and rechristened the name to Department of Geography where he envisioned creating a conducive atmosphere for higher academic standards. As chairman Chandra Pal wanted to build a modern infrastructure and provide a host of facilities for the benefit of students. That the future of geography was being throttled without information technology was his core concern. A deeper remorse was that students, the torchbearers of the discipline, were not being given updated insights into the subject. Reaching out into the twentieth century was not his sole preoccupation. History too fell into his lap.

A moment in history

Shortly after independence, a band of visionaries led by Professor V.K.R.V. Rao and supported by Prime Minister Jawahar Lal Nehru wanted to create a centre for advanced learning and research of high repute in the social sciences and thus the Delhi School of Economics was founded in 1949. Chandra Pal took over as chairman of the Department of Geography in 1999. This was the year Amartya Sen, who had been professor in the Department of Economics at DSchool from 1963–1971 was conferred the honour of Nobel laureate. This was also the year of the Golden Jubilee celebrations of the Delhi School of Economics. It is under the umbrella of the school that the Department of Geography was created in 1959. The Department of Geography acquired a separate building only in 1960. The makeshift cartographic-cum-teaching lab was set up in the present office of the school's director. Three chairmen of the Department of Geography have been directors of the school: George Kurian (1962–1965), Prakasa Rao (1970–1972), and R. Ramachandran (1977–1979). Chandra Pal was keen to give his best to the 50th year celebrations.

For the Department of Geography, it became a moment in history. Various committees in charge of alumni, fundraising, registration, reception, and exhibitions were set up three months in advance of 14 November – the founders' day of the school. Chandra Pal conferred on me the charge of setting up an exhibition. With a band of enthusiastic students, records were searched, slides converted to photographs, theses, dissertations and innumerable research papers and books by the alumni of the department amassed. Joining the celebrations R. Ramachandran came all the way from his retirement retreat at Coimbatore adding a befitting glow to the event. He remarked it was truly a Reminiscence – the caption of the exhibition – and the meet had matched the theme for the day. Chandra Pal had contacted over 300 alumni of the department across the globe and created a meticulous record. While the spotlight no doubt was to be on the presence of Professor Amartya Sen, the Department of Geography had played host to a rich number of students and faculty. A landmark event, it speaks of the talent with which Chandra Pal enrolled most of the faculty and all the students into the department's work.

To Sir, with love

Using a mix of theory, concepts and everyday examples, Chandra Pal navigated class lectures in a way which held the attention of students. While his razor-sharp mind enabled him to answer questions, it was his personality which encouraged student interaction. He liked to teach and encouraged his students to experiment with innovative ideas and ways. That he could train the reticent me into using computers in less than a few weeks speaks of his skill as a teacher. But he was no hand holder; he expected determination and encouraged independence. The chair of the Head pleased him and he enjoyed giving access to students and lending an ear to their troubles. Streaming unhesitatingly into his chamber for advice, references and signatures Chandra Pal's ability to give individual attention to each student won him their deep loyalty. That he enjoyed being with students can be seen in the numerous field survey trips he willingly offered to undertake. Lending an air of congeniality and care was his wife Indira, who accompanied him on numerous academic endeavours – ensuring that geography and students were not separated from his personal life. The family-like feeling he kindled for the department can be seen in the numerous projects students willingly proposed and in which they engaged themselves. It is the pool of alumni and student donations which helped augment the niggardly department funds refurbishing some basic needs. Sound system for lectures, writing boards and accessories such as lecterns, curtains, wall clocks, dustbins for classrooms, overhead lights on tracing tables of the cartographic lab, extended grilles for safety on the main gate, two computers, a printer, scanner and Internet access – all were created for the students. For the faculty, an air conditioner, refrigerator and a tea club were added to the meeting room. No big deal – but thoughtful beyond doubt!

Students were Chandra Pal's top priority and it is ironic that it was they rather than his own sons or a family member who shouldered the gasping Chandra Pal to seek medical help at the University Health Center. Finding the Health Center without any doctor on duty in the middle of the day they hurried him to the nearby Hindu Rao Hospital where he died in their arms. This led to a furious agitation among his students and they lodged a written complaint with the university authorities about the dismal functioning of the University Health Center. Much beyond that it was an act to, "Sir with love."

Leadership – A type

During times when universities across India faced severe budgetary cuts and lax work ethics Chandra Pal set out to find funds and demand discipline. To establish a basic infrastructure for training students to analyse geographical information systems and interpret satellite imageries was his guiding motto. Finances were a major hindrance to install a modern, well-equipped laboratory. His roster of official requests to the following authorities signals his vigour to hunt for funds:

27 July 1999	Director Delhi School of Economics
21 August 1999	Department of Science and Technology
08 March 2000	Vice Chancellor, Universityw of Delhi
26 May 2000	Pro-Vice Chancellor for Development Fund
03 April 2000	National Informatics Centre
25 April 2000	University Grants Commission
31 July 2000	Deputy Finance Officer, University of Delhi

Elaborate research and teaching objectives backed his every effort of reminders, telephones and personal meetings. At the helm of affairs of the Department of Geography he wanted not just the funds but also discipline. An imperative time manager, he simply could not tolerate delays, to him Indian Stretch Time was not acceptable. His punctuality for class, signing out paperwork in the office and holding timely meetings are examples of the same. On 1 January 2000 he took a flight back from Nagpur where he had attended the annual congress of the National Association of Geographers, India. Reaching the department by mid-afternoon and finding the entire staff missing he recorded them absent. It was a cold, foggy day indeed – the chill was to last for a long eight months. Almost the entire staff took a transfer. No personal assistants would pluck up the courage to work under his disciplinarian rule. The result was that for eight months he himself worked at the computer to finish work on time. And there was a lot of work. Never in the past years had the department conferred eight doctorates in a year. Chandra Pal was serving a term as president of the Association of Geographers, Delhi, and joint-secretary of the National Association of Geographers, India. These may seem mere posts but keying details of 1,500 odd members of the latter association into an Excel worksheet is toil. Being a member of the governing body of Kirori Mal College and the academic council of the University of Delhi meant additional meetings. Revision of the syllabus was on the cards and discarding a lot of archaic machines and obsolete equipment which had collected over years in the department was a chore he also initiated. Teaching two full courses and guiding research were added commitments. As chairman of the department, there was much he wanted to get done. As a personality Chandra Pal belonged to the A type – the achievers. His leaps from Shamili to London, academics to industry, and back to academics and from agriculture to electoral geography to political geography of India – all signal his drive to achieve the best for himself and for the Department of Geography. In a brief 16 months as chairman of the department he had tried to achieve too much. The load of work, hunt for funds and projects floated coupled with the frustration of delays created an unnoticeable syndrome of stress. While his mind was energetic – his body could not keep pace and recorded the brunt. A massive heart attack took him in a matter of a few seconds. Pushing the department towards a new century he did not live to see 2001. He had said it was a last season.

A last season

On the side lawns of the faculty block of the Department of Geography grows the *Salmatia Malbarika*, a tree locally called *Semal*. Its carpels hold brilliant red petals. Belonging to the group of angiosperms, the wind disperses its seeds. It is not clear if the tree existed before the building or anchored later. Whatever the sequence today the tree grows a bit too close to the wall of the department building. Flowering twice a year, in March and August, its red flowers carpet the lawns and crown the building of the Department of Geography. For those on campus it is an ornamental delight. Towering to a height of over 15 m is an indication that its strong roots have dug deep and struck at the foundations of the building. Chandra Pal wrote, "the roots of trees have gone deep into the foundation and branches are covering the roof, [the] foliage, causing seepage in the walls." (Letter to the chairman of Garden Committee, Department of Botany, 15 September 1999.) Two deforesters came to hack the tree. Environmentalists like me at times can be extremists. An argument brewed between Chandra Pal and me and the matter was left pending. I could see that despite the apparent beauty, the *Semal* was damaging the structure of the building but yet I pleaded. A concession was granted for a year and he said, "I shall not see the tree the next season." That was March 2000. Today, March 2001 as I write these words, he is no more, he did not live to see the tree again – it was a last season.

In the overcrowded schedule of administrative work, teaching and meetings, the crystallization of his ideas got delayed – this time forever.

Crystallization of ideas

Possessing creative ideas is distinctly different from having an identity. One can have ideas but no identity; conversely, there can be an identity without ideas. The possibility of gaining both – idea and identity is also not uncommon. The fact, however, remains the two are separate. Chandra Pal established an identity as a political geographer prior to the in-depth crystallization of creative ideas on this theme. While the embryonic seeds were sown with the two research projects in the 6th and the 7th parliamentary elections back in 1978, his identity as a political geographer was well cemented by 1990. A serious crystallization of ideas gripped Chandra Pal between 1991 and 2000. How ideas brew and evolve is chemistry, which I am unable to fathom; indeed, in most of us, they take time to mature. The break in Chandra Pal's creative line of thought was more due to lack of opportunities than procrastination of ideas.

The changes in the syllabus of Political Geography, a course which Chandra Pal had introduced in the department in 1972 – voices a silent but sure morphogenesis of ideas. Take the title and contents of the course called Political Geography prior to 1983, it carried only one section on India but by 1993 a clear bifurcation of ideas had taken place with two courses, namely, Political Geography and Political Geography of India. From the theoretical and global treatment of the theme attempts of Indianisation were well under way. Whereas earlier courses leaned much towards an emphasis on elections and voting the latter had diversified into areas of regionalism, environment and the politics and problems of nation-building and international relations. In tune with the course, the design of his book was making headway.

Strewn among his heap of papers is a note carrying the title he first thought of for his book –Politics and Development, dated 1991. In a letter from 1992, he expresses,

If I can be relieved from my normal duties and routines here, I can surely put forward these ideas in the form which can become the basis of my future book. I am at such a ripe stage that I need only eight weeks to complete the job.

(Letter to Professor D.B. Knight, 22 February 1992.)

The weeks never came. In fact, thereafter, there is a virtual lull with no reference towards a book until 1997, the year he took a sabbatical. By the time of his return from study leave he had a well-structured title – *A Geography of the Indian State* which was changed in 1999 to *Political Geography of Contemporary India*. The blueprint of his book had been cast. He needed time to pen its contents.

That a more thorough and systematic research investigation was underway compared with the decade of the 1980s can be seen in the nature of papers and public lectures he delivered throughout the 1990s, "Darwinism in New Global Order" (Washington, 1992), "Power Structure in Indian in Democracy" (Delhi, 1993), "The Asia-Pacific and Global Geopolitical Change" (Tokyo, 1993), "Environmental, Politics and Status of Women" (Dhaka, 1994), "Constitution Delimitation Policy in India" (Chandigarh, 1996), "Vanishing Borders: The New International Order of the 21st Century" (Kuala Lumpur, 1996), "Geographical Factors in Political History of India" (Rohtak, 1997), "Factors in the Political History of India" (Syracuse University, New York, 1997), "Geographical and Political History of India" (Chandigarh, 1998), "1998 Elections: Geographical Analysis and Predictions" (Delhi, 1998), "A Geographical Analysis of the Background of the 12th Parliamentary Elections in India" (Delhi, 1998), "India and the New Geopolitical World Order" and "Assembly Elections: 1998 Causes and Consequences" (Chandigarh, 1998). Editing the book, "Readings in Political Geography," (1994) and research articles like, "India ASEAN and the New Geopolitical World Order" (1998) and, "A Century of Constituency Delimitation in India" (2000) speak of the fact that identity and ideas had merged. He had come full circle.

Compared to the 1980s when carving an identity was the more dominant agenda, the works of the 1990s reflect a definite drive to comprehend the social and historical factors which were instrumental in shaping India's political landscape. Close scrutiny of his notes depicts an earnest attempt to understand the historical dimensions which would augur well for the analysis of political processes and patterns. Painstaking study of the break-up of all political parties at the national and state levels in India, tabulation and analysis of data from the first general election in 1952 to the 13th election in 1999 give glimpses of his scientific vigour. He had drafted the issues of regionalism in India right up to the creation of the three new states of Jharkhand, Uttarakhand and Chhattisgarh. Incorporating the influence of politics on the environment, especially environmental movements, he highlights laws and policies that show his in-depth ideas. An abstract of his book which he let me have a few days before his demise best sums up the crystallization of his ideas:

The political geography of India, and for that matter of any country, is the outcome of complex processes over time. The present cannot be understood without the understanding of the past. For instance, the schism between India and Pakistan, the churning in the Hindu society with the upward mobility of many castes, the regional variations in sub-cultures and politics etc. cannot be comprehended without taking the varied experiences of numerous localities and areas of society, economy, polity under different princes, kings, emperors, feudal landlords, etc. at different times along with the spread of different indigenous and foreign religions. The spatial differences in society, demography, culture, resources and economy over a physical base, and political structures and processes, and their interrelationships and interactions over time are reflected in the political geography of India. This book on the Political Geography of Contemporary India attempts to provide a comprehensive treatise on the above factors.

I know the book would have been a delight to read.

Innumerable notes under the general heading, "Book Support Material" can be classified into diverse subtitles of administration, militancy, globalization, geography and politics, political parties, society, history and archaeology, technology, strategy and power. Ideas get sharpened as readings are sifted and categorized. That the craft of geography had been mastered is all evident.

The last half of the decade of the 1990s was perhaps the most prolific of his life. It was a creative decade for him. A sense of frustration and sadness does grip when one sees the exhaustive bibliography, several notes and articles, facts and data logged faithfully in his laptop with many a website awaiting a browse. An era of the trio of Professor R. Ramachandran, S.G. Burman, and Chandra Pal Singh in the Department of Geography, Delhi School of Economics comes to an end – a new one unfolds. While projects and infrastructure in the department could be pipelined and reinitiated in administrative routine, the crystal ideas have had to die with Chandra Pal.

Acknowledgements

I am grateful to my research students, Deeptima and Roshani, for their support without whom a lot of checks could not have been made. I appreciate that in spite of heavy hearts, Chandra Pal's children and wife Indira, all in the US, all took time out and allowed me to interview them via email correspondence.

Notes

1 Mackinder, J. Halford, "The Geographical Pivot of History," *Geographical Journal* 23(4) (1904): 421–437.
2 Wren, P.C. and Martin, H., *High School Grammar and English Composition* (Bombay Maneckji Cooper Education Trust, 1936).

4 Mapping Gopal Krishan

As a geographer in India, Gopal Krishan (GK), did not lead an exploration or expedition, nor did he build a school of thought; establish a new department or float a journal. GK as a geographer in India did not even live the dullness of rotting routine, recycled notes, bored lectures and disparate seminars. The roster of minimum needs, so typical to life in the universities of India today, has not been the charter of his work.

What then did GK as a geographer in India do? What is the map of the life and works of GK? Where does one begin? Compiling a pool of data, facts and literature, along with ground verification can be the start towards creating a map. Official correspondence, diaries, travel documents publications and manuscripts were all accessed to provide fillers and footnotes. When treading sensitive turf, interviews helped sift controversies. A rich fieldwork enables a map to emerge.

Mapping GK (1940–2002) was entrusted to me in January 2003, six months after he retired from the portals of academic life. I accepted, for my credo is that mapping the other leads to mapping the self. This is the reason and dividend.

Map 2002: A Snapshot A cornucopia of works embellish GK's Map 2002. Half a dozen places of formal education, over a hundred research publications, a score of books and projects, several foreign visits, some news reports and reviews, outstanding scholarships and honours, numerous official and academic assignments form the landmarks of 2002. A technique in mapping is to plot each of these works on acetate transparencies and sift them one on top of another, like the layers of icing on a cake. What emerges is a complex picture. The ensemble is not a monochrome. The palate of GK's research is filled with diverse themes, regions and methodologies. Medleys of teaching, research, consultation and administration are markers in the file of his work. Dense lines traverse between academics and administration, universities and research institutes in India, and the Western world. Honours, scholarships and prestigious assignments demand their space of recognition. In the metaphor of this map, the colour, line and points are icons of GK's life and work. Constructing a graphic for a life such as his invites deep thought. What is the size of the map? What influences bond to give it shape? What are the gradients of his academic landscape? What is the appropriate name for the map? Mapping GK involves this and more.

Size: How can one measure the size of academic work? If all of GK's printed works are charted into three categories: the word, the map and the table, then ten octavo volumes of 400 pages each would contain the words; an atlas of 300 pages would compile the maps, and one statistical volume would list the 700 tables. If all of GK's manuscripts drafts, supervised research, collection of notes and references, books and journals were stacked, a good number of bookracks would fill to the brim. As an academician is GK only the sum of his print material? How does one access the role of an academician as a thinker, teacher, administrator and an informed citizen?

The history of cartography confirms that accurate measures came only when advanced instruments were invented. For centuries maps were not made to scale. Diagrams, sketches and pictographs sufficed in more ways than one. Taking this as a cue, let us move beyond pinning size towards the understanding of an immeasurable yet significant territory. What is he like?

Coordinates: Landlocked between friends and foes is not the territorial configuration of GK; neither is a lofty arrogance that looks down at the plains. Similarly, the plains looking up for favours towards the higher echelons in the profession is also not his way. A wishy-washy delta, where rivers channel in directionless ways is surely not GK's terrain. Most confirm that GK is a recluse. Some see mannerisms similar to the secretive Phoenicians while others testify to an overt sensitivity. My first impression was a face neither open, nor secretive, but a remote face, of a person with a purpose and philosophy. Shy and introverted, those who know GK agree that especially about himself he says little. He is an island unto himself.

Genesis: On 5 July 1940, in Bhalowal, now a town in the Sargodha district of Pakistan, Gopal Krishan was born. Located on the Punjab plains, 400 km west of Amritsar, the town was famous for citrus grading, textile spinning and cotton ginning industries. While his father pursued a graduate degree at the University of Lahore, GK spent his infant years with his grandfather, who was a manager in a cotton-pressing mill. About his childhood, GK recollects that disciplinary duty and devotion towards the Divine were common family discussions. Being ardent devotees of Lord Krishna explains how GK's parents came to name the eldest of their seven children, Gopal Krishan, synonyms of the names for the same deity. A miniature temple in GK's study confirms that even today his faith is ever-abiding.

GK's father, Shri Kharaiti Lal, an officer of impeccable integrity, retired as Deputy Director of Industries, Punjab Government. When GK was growing up life was hard, the Indian subcontinent had split, but the family were fit and survived handsomely and with dignity. While a pious outlook, durable character and commitment to duty are GK's genetic loads where did he acquire his intellectual bent of mind? A genealogy of ten generations yokes the family to the Vale of Kashmir and the intellectual *pundits*. The possibility of a lineage cannot be overruled. Though Punjabi Brahmins, yet some of Kharaiti Lal's official correspondence signatures the surname *pundit*. The Caucasoid complexion, light eyes and stout build of GK and his family resembles the race of the Vale.

Perhaps it is the ecology of a devout family, the time of political unrest, and the mores of the eldest son, along with an alert cerebral sensitivity that metamorphosed GK, a withdrawn child, to omnivorous reading, reticent to the world, an island unto him. In a world of his own, in his own way GK tried to excel. Crossing strata of different time and work frames of different kinds, GK's desire was a pursuit for the best. His was not the desire of a wishful thinker. Backed by a surge of energy and a spirit of enthusiasm GK moved from one step to another, one project to another, one research paper to another, never to stop, look back or take stock. The contours of learning honed GK's ability to excel.

Contours of Learning: Shifting from Lahore to Amritsar to Kalka, a chequered primary education gained stability only when GK was eight years old. On a ridge of the Middle Himalayas, the British *raj* built their summer capital at Shimla. The Sir Frank Noyce School, nestled at Phagli, Shimla was a transitional base. In 1950, GK was admitted to the puritanical Dayanand Anglo Vedic School (DAV College) in the capital. Its tin roof and wooden benches, which personified stern discipline, did not thwart the 10-year-old lad's joy of schooling. Nostalgic, he recalls, "…School days were happy days, in the lap of nature, in a reverie of thought and study, I most enjoyed organising class boys for different activities." Being nominated as the class prefect, winning declamation contests and scholarships, GK was an ideal student. A good school education was a vantage point from which GK could focus on knowledge rather than mundane gains. Ingrained are memories of the Grand Hotel, Shimla. Seated amidst the elite, when Vijay Laxmi Pandit, the younger sister of then Prime Minister, Nehru, announced the awards for a declamation contest, the first prize went to GK. The theme on which the young teenager spoke was the Glory that is India. It was a glorious day. Not a sense of victory but the felicity he gained from excelling flexed him to work. The significance of this early hard work cannot be overlooked. It gave GK the predisposition and stamina to simultaneously pursue a number of goals. A most useful arsenal for times ahead.

After a brief study at the Sanatan Dharam College, Shimla and graduation from DAV College, Jalandhar, GK selected geography for his master's degree. With his parents in Chandigarh, the Department of Geography, Panjab University was geographically the obvious choice. In the pre-1960s the history of Panjab University bore a resemblance to the itinerant early years of GK.

The British founded the Panjab University at Lahore in 1882. It was the fourth teaching and affiliating university on the Indian subcontinent. With the division of Punjab in 1947, Lahore, the old capital of united Punjab, was lost to Pakistan. There was need for a capital to house both the post-Partition government and educational facilities of the Indian counterpart of Punjab. Before Chandigarh was formally inaugurated in 1953, the university resembled a vagabond. Its administrative office was located at Solan, while departments were scattered in different towns. Government College, Ludhiana was the refuge of the Department of Geography. In 1956, Panjab University finally dropped anchor at Chandigarh. A palatial second floor, atop the Department of Geology was ready for occupation by the Department of Geography in 1960. A nature's location as below the earth lies the rocks. For nearly 13 years the department had wandered form Lahore, Ludhiana to Chandigarh. When it opened office for admission, GK was the first among the first to enrol.

Geography was a subject GK relished. Those were days when remote cursors to draw maps on computer screens were unheard. With pencil and pen, "...coloring maps and softening the unevenness of the shades of blue and brown with swabs of cotton was as much of a delight as the philatelic hobby to classify countries." The range of knowledge geography embraced was an added asset. The post-graduation report card placed GK with a first class first in geography. Despite an inherent aptitude for the subject, GK did not want to be a geographer. Academia was not his first choice. His heart was set neither on research nor teaching. GK's ambition was to serve the Indian Administrative Service. Few will deny that even today students, parents and society hold recruitment with the services in high esteem. GK was no exception, but fate had other plans.

Turn of Fate: Teething problems gripped the Department of Geography, Panjab University in its infancy. G.S. Gosal followed as the chairperson. He nurtured a vision to model the department along the lines of his alma mater (for doctorate), the University of Wisconsin, America. The mission called for loyal comrades, but this became its Achilles heel. P.K. Sarkar, a doctorate from Queen Mary College, London joined the Department of Geography in 1960. In around a

Gopal Krishan in Department of Geography, Panjab University

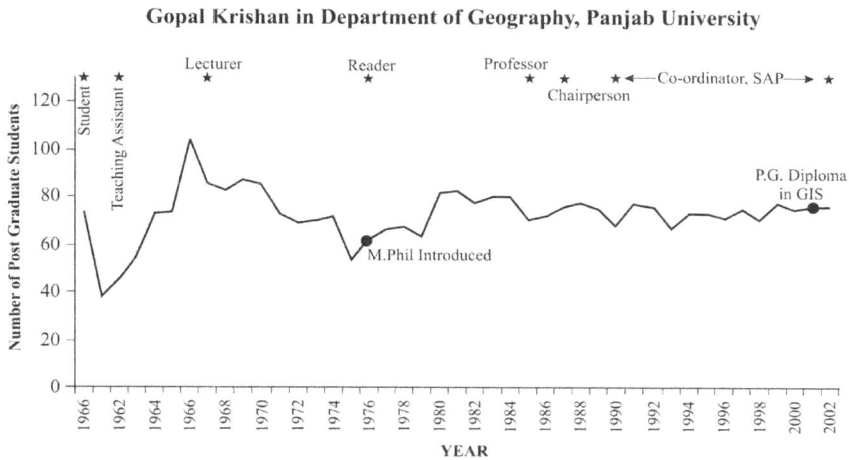

Based on the annual reports of Panjab University

Figure 4.1 Gopal Krishan at the Department of Geography, Panjab University.

year he left for the University of Nigeria, Nsukka in Africa. Meanwhile, Amrit Lal left to join the Delhi School of Economics, University of Delhi in September 1962. There was J.C. Sen, but he chose to confine teaching to his specialization in mathematical geography. The department was acutely short of faculty. Migrations and shortage of skilled staff made it difficult to run a postgraduate programme offering 12 courses. Gosal was quick to recognize in GK an intellect for research and a temperament to commit. He persuaded GK to join the Department of Geography. (Figure 4.1)

As soon as the post-graduation results were announced GK joined the department as an instructor in 1962 and in 1964 was elevated to the position of lecturer. As if in abeyance to fate, GK sacrificed his longing to be in the services, Gosal went all out to care for the prized student with GK becoming his first doctoral student in 1968. Evidence of this labour of love of both student and teacher can be found in the laudatory reports by the examiners, P. Simkins from Penn State University and N. Thrower from the University of California. While GK proved his mettle, Gosal's philosophy for automatic justice ensured multiple increments accompanied his student's ascent from instructor in 1962 to professor in 1985. Though GK was Professor of Regional Planning at the National Institute of Urban Affairs, Delhi in 1983; a visiting professor at the University of Bremen, Germany in 1993; and the University of Pecs, Hungary in 1995; it is the Department of Geography, Panjab University, Chandigarh, which was the permanent address for the island. Here numerous interactions hewed GK's terrain of thought and work.

Enduring interactions

Influences near and far enriched the shores of GK's academic life. His verve ensured continuity.

Outshining teacher

Coming into our lives at an impressionable age, teachers leave an indelible imprint. Opting for a profession in academics reinforces their role, which at times is difficult to wriggle out of. Among a number of teachers who have influenced GK, all pale before Gosal. The erudite Gosal carried an air of confidence and stability, which invoked GK's high regard for him. The virtues of clarity, synthesis of ideas and organization are nuggets from a great teacher. Tradition abhors comparison between teacher and student but parallels help gauge influence. Gosal devoted 24 years to the department, GK spent 40; Gosal was chairperson for 20 years, GK was at the helm of affairs for 15 years. Gosal spent over two years in the USA; GK spent almost two years in Europe. Gosal started a diploma in cartography in 1963, GK initiated a postgraduate diploma in geographic information systems in 2001; Gosal compiled a gazetteer for the city of Chandigarh, GK drew up an atlas of the city in 1999; Gosal was the first president of the National Association of Geographers, India, GK was its 22nd president. As president, Gosal announced his concern for the discipline; GK supported the publication of *Voice of Concern*. Gosal was the first to introduce a course on population geography in India, GK was the first to introduce a course on administrative geography. Both did much good in their own different ways. Scientific progress would stultify if we trained our students to be truly faithful disciples. Change endures research. In the pool of interactions teachers draw pride when students outshine. While the academic parent has pay-off, the desire to harness potentials of the self took GK time and again to the West.

To the West

In bits and pieces, for durations both short and long, as a young scientist or visiting faculty, in ten visits GK spent nearly two years in Western universities. His passport endorses five entry permits in 1974, 1975, 1981, 1982 and 1993 to the UK, four visas for Europe in 1993,1995, 1996 and 1999 and two seals of the immigration authorities in the US – one in 1989 and the other in 2002. Universities big and small, some steeped in history, others making history, provided GK a myriad experiences. The universities that GK visited are like a nested hierarchy of small to large towns. On one end of the spectrum is the Bemidji State University and University of Cape

Girardeau with a student strength of less than 5,000. In the middle range fall the University of Pecs, Hungary, University of Bremen, Germany and the University of Liverpool, with 15,000 to 20,000 students. The University of Cambridge is the largest among those which GK visited. Each share a different birthday: Bremen is the most recent, being established in 1971, Bemidji was founded in 1919, Pecs, dated 1367, is the oldest university in Hungary, and when GK visited Cambridge, the university was celebrating its 765th anniversary.

The culture in universities, differing in size and age, was a window to launch contrasting contacts. The exchange programme sponsored by the University Grants Commission, India gave GK the opportunity to spend 100 days in the UK. GK was 34 years of age and this was his maiden trip to a foreign land. A panoramic view of university life unfolded as GK visited the universities of Cambridge, Oxford, London, Cardiff, Swansea, Birmingham, Durham, York, Manchester, Edinburgh and Glasgow while based at Liverpool. These visits whet his appetite and in less than 10 months he had returned. At the invitation of the sponsors from the UK, he came to attend the Indo-British Seminar at Cambridge in 1975. In the throes of discussing rural–urban relations, the seminar's focal theme, GK was struck by the ambience of the venue. And why not? It was the room in St Johns College where Wordsworth, poet laureate at Cambridge in 1791, wrote the famous Daffodils! While his first visit to UK was an eye opener the next was a mind opener.

The Royal Geographical Society Award gave GK a chance to pursue research at the twin centres in Cambridge: The Department of Geography and the Center for South Asian Studies. GK was delighted with the arrangement. The department at Cambridge headed by an internationally known scholar Michael Chisholm and the centre under the directorship of B.H. Farmer became, "…where one would meet the most gifted scholars of South Asia not only from Cambridge but from all over the world." GK was here for a period of six months.

While Chisholm ploughed GK's grasp of administrative area reforms and structures, Farmer spruced up his inclination towards problems of development in India. Commissioned with the Royal Engineer Services, Farmer had had the opportunity to work in Singapore, Sri Lanka and India and gain an insight into the agricultural colonization of South Asia. When GK was visiting Cambridge, Farmer was finalizing his book on South Asia. These works inspired GK.

Hailing from a country bisected by the Tropic of Cancer, at the University of Liverpool, GK was sure to be the most tropical minded department of geography in the UK. Here the research of scholars such as M. Prothero and R. Steel on epidemics in tropical Africa, migrants and malaria and new interpretations of census data became themes for sharing. Prothero's congenial intellectualism and personal affection created greater warmth. An intellectual bonanza was GK's interludes with Haggerstrand, Taylor, Haggett, Bradnock and Chapman, among others. While research in the

UK brought home the sagging status of Indian geography, contact with the Indian diaspora in America kindled hope. Scholars like A. Dutt, S. Bharadwaj, R. Tirtha, and Harbans Singh kept the spirit of the geography of India alive in America and lent support to scholars from their native country. This was apparent at the University of Akron and Kent State University. Igniting GK with innovative themes for research in India were scholars like Dickason at the Western Michigan State University.

Novel and perhaps more satisfying was GK's interaction with the European mainland. The reason is not hard to find. The UK was the country where GK spent time on research. Europe was a destination to lecture, a platform to interact and exchange views. Furthermore, whereas GK had read the works of geographers he met in the UK and America, the same was not true for Europe. It was a new world of novel experiences. While GK travelled across Germany, Netherlands, Hungary, Austria and Slovakia, he stopped to lecture at many places, yet two universities, the University of Bremen, Germany and the University of Pecs, Hungary left a lasting impression.

It is in Pune that G. Bahrenberg and GK met for the first time in 1991. Poona and Bremen are twins. The twin city concept draws from a mutual exchange of scholars and other interactions between a pair of cities. The common interests of Barhenberg and GK on regionalism and territorial knowledge set them discussing. In 1993 GK was invited to the University of Bremen which has a strong orientation for interdisciplinary research. GK had to be on his toes as issues of comparative regional and urban development became the theme for a series of lectures. While in Germany, GK was confined to a few scholars, but in Hungary the entire country won GK's heart.

Albizia lebbeck

At the entrance of the botanical garden in the University of Pecs, stands the 5 m Albizia Lebbeck. Within a schema of temperate species, this tropical tree is hard to miss. The grey black bark, peeping through a tangle of oblong green leaves, lends conspicuous beauty. Indigenous to the Indian subcontinent where it is popular as *siris*, the *one* in Pecs seems acclimatized to Hungary with its warm summers and cold winters.

On 15 November 1996, the day Hungary celebrated its millennium amidst a standing ovation from the senate, GK rose to humbly speak: "...There is a moment in the life of an individual when one feels that there is nothing more to aspire, by conferring the Honoris Causa Doctoral degree, you have honored my country, my university, my department." It was a kind of epiphany for GK's quest to excel.

Reaching out through interpreters, GK lectured across the nation. Audiences of the Hungarian Academy of Sciences at Budapest, Debrecen, Szeged, Bekescsaba, Nyiregyhaza, and Kecskemet perhaps for the first time met a scholar from India who revealed to them what India really is. Articulating issues of cultural pluralism, democratic secularism, problems of development and how India stayed integrated, GK took special courses on India and urban development. He read among the Hungarian students a passion to understand India. His lucid presentation so stirred their imagination that some came to India as tourists, while others availed grants under the Indo-Hungarian exchange programme. Their sentiment sustains: "...GK is a unique geographer for Hungarian students, he could bring home [the] basics and beauties of India... he is the most important reason for so many admirers of India in Pecs and Hungary," is a recent comment in an email from Zoltan Hajdu, senior fellow at the Hungarian Academy of Sciences, Pecs. If all scholars were ambassadors like GK, many bridges between India and other nations would be built.

Planted by GK, the roots of the lebbeck trace back to 1991, when Jozsef Toth, present rector at the University of Pecs visited Panjab University. A scholarly bent and an inclination for building institutions gelled the two from the day of their contact. GK was in Hungary in 1995, 1996 and 1999. The lebbeck personifies the life force between Indians and Hungarians.

Like fleeting birds with the passage of time the West left fond memories, but it was in India that GK built an edifice for the advancement of the discipline to which he was committed.

On home turf

In varying capacities, GK has been to one-third of the 78 departments of geography in India. But an official visit is neither his sphere nor style of influence. It is through the channel of publishing over 40 research articles in ten leading geographical journals of India and relating through sagacious scholarship that GK's fame travelled across India. GK has been speaker at both the R.N. Dubey and S. Muzaffar Ali memorial lectures. The National Association of Geographers, India is the largest society of geographers in India. The Indian National Cartographic Association is the largest mapping society of India. GK has been president of both. All this places him in the front rank of geographers of the country. Modest to dismiss ranks as military and not academic, today he says, "... not the services, nor the west, but Indian geography for me has been the best!" Taking pride in his subject, GK was not a prisoner to his discipline. To travel wide he not only crossed national boundaries but also stepped out from the confines of the discipline.

Transgressing boundaries

Aware that academic compartments are artificial and limiting, a range of social scientists such as G.K. Chadha, Kulwant Singh, O.P. Mathur, A.C. Jhulka, Victor D'Souza, Ashish Bose, A. Kundu, and Kulbir Singh were constant supports. From professionals like town planners, architects, policy experts, and administrators across the rungs of the central and state services GK learnt to walk the tightrope between pragmatism and idealism. Over 25 publications by GK have a place in non-geographical journals. The contribution of economists, sociologists, and planners as authors within the ambit of this book is proof that for GK knowledge is vast and seamless. Forging working partnerships, a string of premier institutes knit themselves into GK's academic life. Exhibiting an ease of adaptability to new situations and groups, GK's zeal produced works that set milestones of continuity with cross-disciplinary bodies.

With an invitation to a seminar, GK first stepped into the National Institute of Urban Affairs, New Delhi in 1981. In 1983 he was appointed Professor of Regional Planning at the institute. A number of projects including the national capital region plan, the slums of Siliguri and financing municipal services, owe allegiance to this office. After a brief two years of GK's presence, the director, O.P. Mathur, expressed, "... GK has made [a] significant contribution to the institute's research activities and helped in the establishment of a doctorate programme, he has assisted us in editing the institute's journal." Though GK returned to university life, his bonds with consultancy remained. A recent query from Mathur reiterates the sentiment, "...his academic contributions helped the National Institute of Urban Affairs to grow and develop into a quality research institute."

GK has the flair to convert even the relatively drab environs of an administrative forum into a venue for cross-disciplinary discussion. The main role of any academic staff college is to organize orientation and refresher programmes for the teaching community. India has 45 such colleges. GK was appointed Honorary Director, Staff College, Panjab University in 2000. An administrative post became yet another conduit for interdisciplinary learning. Many of GK's official speeches carry smatterings of disciplines across the board. His dexterity to administer an organization like a bureaucrat, but run it like a scholar, is perhaps the reason why the vice chancellor insisted on a second term of directorship for GK.

In 1981, the Center for Research in Rural and Industrial Development, Chandigarh, invited GK to be a member in a border area development project. This set rolling two decades of invitations culminating in GK's appointment in 2002 as senior professor and honorary director at the Population Research Centre, the very same place Rashpal Malhotra, the founder director was emphatic that, "...GK is here, less as a geographer and more as a scholar whose vision straddles numerous disciplines." In a tone seasoned with experience he adds, "...the world today needs minds that do not disintegrate but synthesize." GK is a generalist in an age of specialization. If ever at a loss where geography in India can be spearheaded, GK's curriculum vitae is a well-tested list. If ever at a loss from where GK draws his unflinching strength, visit his home.

Freed to interact

GK draws energy from his tastefully furnished home sparkling with crystal and care. Sacrificing a lucrative teaching career his dutiful wife, Kanti, single-handedly shouldered the chores of domestic life and reared two accomplished children. There is little doubt though that GK instilled in them the values of professionalism very well, due to which Mani and Gaurav stand well placed as a doctor and engineer, respectively, in America. Thus freed from most tasks that go with running a home, GK was able to pursue academics to bring home laurels to share with the family he cares for and those he loves including Amit, his son-in-law and Sahiba, his grandchild! Such is love that the sacrifice of many gives the freedom that builds the success of one. GK embraced a family larger than the nucleus unit.

Concerned treasure

Sky television and national papers carried interviews and photographs of GK in Hungary as he was placed on a pedestal by bestowing the degree Doctor of Science. Achievements are never small or big; occasions to excel each deserves due recognition and celebration. Returning home, while family and friends rendered heartfelt congratulations it was GK's students who gathered to share and shower their appreciation. If students make teachers proud, the other way round is equally true.

Tailoring recommendations and helping students to fill in applications placed many of GK's students in foreign universities. While it is the duty of teachers to pen confidential reports and help students further their aspirations of higher education, the other way round does not hold true. Bon voyage is not the only feeling that GK's students carry as they set to sail. Let us take Anamika and Satish who share three commonalties: both were GK's students at the University of Panjab, both are in the US and both have been instrumental in inviting GK. Proud to show off their teacher from India to his American counterpart, in 1989 Satish invited GK to the Minnesota State University of Bemidji, while Anamika organized GK's visit to the Southeast Missouri State University in 2002. While some students took GK out and abroad, a few plunged him into the pathos of inward reflection. A tragic car accident crushed K.N. Dubey on 5 March 1992. He was GK's fourth doctoral student. With ink dipped in grief, GK groped for words, "…to write a foreword is a deeply sensitive assignment… it forced me…to reflect on the meaning of life at the individual level and that of development at the societal level." Families after all are not just biological.

The register of admissions at the Department of Geography, Panjab University verifies that during his tenure GK addressed nearly 1,500 postgraduate and 300 research students. Leafing through timetables from the 1960s to 2002, one learns that he lectured for 25,000 hours or nearly three years of non-stop teaching! These official hours and slots in the classroom are of course no timekeepers of the intense ways and deep moments that go to mould the student–teacher bond.

A carton in GK's study stands out as it bears a one-line request: "Treasure do not destroy." Brimming with greeting cards, thank you notes, letters, email prints, small momentous, photographs, these are GK's most prized possessions. The temptation of sharing a few from this kitty is irresistible. Maintaining the ethics of anonymity here are some, "... your lectures made a world of difference to how I see the world – is that why you say geography is important?" On a greeting card with a watercolour of a small *diya* another student writes, "... from you I learnt to value geography, I value your place in my life and... wish to remain your student forever." Another shares, "some teachers make a difference ...teachers like you." Of GK's 120 publications, 17 per cent are co-authored with students. Names cited on the acknowledgements page confirm that many assisted in the numerous projects that GK managed.

Ever so gentle, ever so kind were GK's ways of shaping the young mind. Never dictating, sermonizing or prophesizing, his students recall that his softly delivered lectures found swift grasp. Rarely burdened with exhaustive references, or bogged down by miles of data, a deep philosophy and special provocation is what made GK's lectures differ from the stereotype. While there is no theory of how one learns, the fact remains that teaching is an art in which GK excelled. To become a missing name on the list of GK's roll call would have been a lifelong remorse had *Concern* not come my way.

When GK was nominated president of the National Association of Geographers, India in 2000, in a quest to deliver something unique, he set before himself the agenda of compiling the presidential addressees of 22 of his predecessors. Snowed under work his hindsight warned that his brainchild faced failure. Destiny chose me as the surrogate. I authored the book and titled it *Voice of Concern*. In the six months of intensive contact learning, GK tamed views, perished faults and nourished creativity. A true teacher plants roots and provides wings. My ability to sketch the biographies of 22 top geographers of India who were presidents of the Association could be what encouraged the editors of this volume to select me for mapping GK. What goes around comes around. Concerned treasure: GK for students and students for GK.

What would GK have been without these interactions? The elegant hand crafted leather scroll and embossed medal on which is inscribed the message of the Doctorate of Science; the fragrance of pine in the air around Lake Itasca, the source of the Mississippi River; the rock promontory at Cape Girardeau; the quaint bridge on the River Cam which lends Cambridge its name; the trail of the Danube as it rises from the Black Forest mountains in Germany and traverses through Austria, Slovakia and Hungary; the polders on the coastal belt of Germany; the visit to Graz, a highly affluent alpine habitat in Austria; the enthused Slovakians with their newly formed republic in 1993; the warmth of Hungarians; the Dutch laughter; the German professionalism; the British reticence; and the American exuberance – the views and viewpoints of all these people and places enriched the island within GK.

If GK had not ventured to interact with the cross-disciplinary realm his research and lectures would be loaded with theory but lack the practical, would skim on others' experiences rather than draw from personal conviction, and the philosophy of his discipline would starve for live application. The teachers, students, home, colleagues and critics were the whirlpool of interaction that sculptured GK's academic landscape.

Academic landscape

Soaring above material pleasures, physical comfort or idyllic leisure, GK built a prolific academic landscape. He had barely been confirmed as faculty in the department when he set out for the field to research how a planned city builds its spatial linkages. At 22 years of age, when most postgraduate students are foggy about their careers, GK was already the author of his first published work.

From then on there was no looking back. Running the fast lane track, in 40 years GK placed nearly 150 of his own works on the shelves that line his study. Eight books, one atlas, 20 projects and 120 research articles. GK's academic portfolio also encloses credit for supervising ten doctorates and 24 Master of Philosophy dissertations. Lecturing is a staple need of the teaching profession. His wealth derives neither from a sweepstake nor a decade of sweat. A remarkable consistency reiterates that for GK lectures, research, and publications are a way of life, in fact they form his life. The graph of the academic landscape is proof of prolificacy. Every year GK published four research papers, each year he adds one new journal as an outlet for his research, every year a different cross-disciplinary body was attached to GK, every second year a project was sponsored to him, every four years there is a foreign visit, and every 5th year a book. The average output is a dizzying gradient of research.

It is generally agreed that the span between being a lecturer and a reader is the time when research is spawned. These professional rungs, however, are neither a signal to toil nor relax. GK's annual productivity of publications from lecturer to reader is 2.0, from reader to professor it climbs to 3.3, after GK becomes professor the output mounts to 5.1 publications. While all years were busy, some years were all the more dizzy. In 1993, for example, GK published ten articles, completed one project, travelled abroad twice and supervised the submission of one doctorate, all this along with a full administrative and teaching load. At times he did burn the candle at both ends. As his publications were brought out by ten foreign and 29 Indian journals and requested in 32 different edited books, GK netted wide repute (Figure 4.2).

A charge often levied on scholars in India is that they add mileage to the number of publications through cloning, joint authorship and substandard writing. GK's stock of 144 publications contains 37 that are co-authored, while 20 are reports of teamwork in a project. Appraising quality is subjective and there is no judge better than an honest self. GK was requested to rank a hundred of his publications on a scale of 1 to 10, with ten being the creative best. His evaluation:

ACADEMIC LANDSCAPE

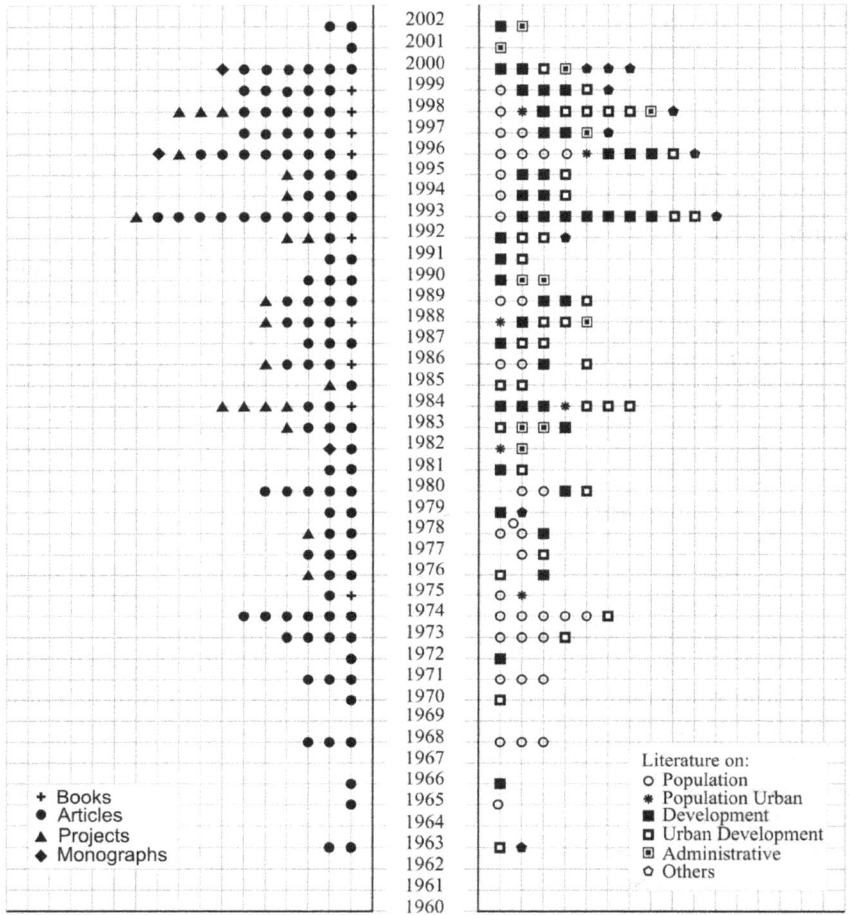

Figure 4.2 Academic Landscape.

35 per cent of his works mustered a score of 5 and below, 34 per cent fetched between 6 and 7, the remaining 31 garnered 7.0 and above. GK's extreme modesty cannot moderate the reality that a scholar's research is a mix of the worse and the best; what counts is to move ahead in search of the very best. Taking a *longue durée* view one is struck by the productivity of his work.

How does GK manage? The best of Indian scholars, especially geographers, are comfortable with the oral tradition. Geographers in India are known for their unwillingness to write, and barely publish one article every three years. Seclusion and detachment, the essentials for writing do not come easy to the social race, but an island is a sanctuary for contemplation and publication.

Research Terrain

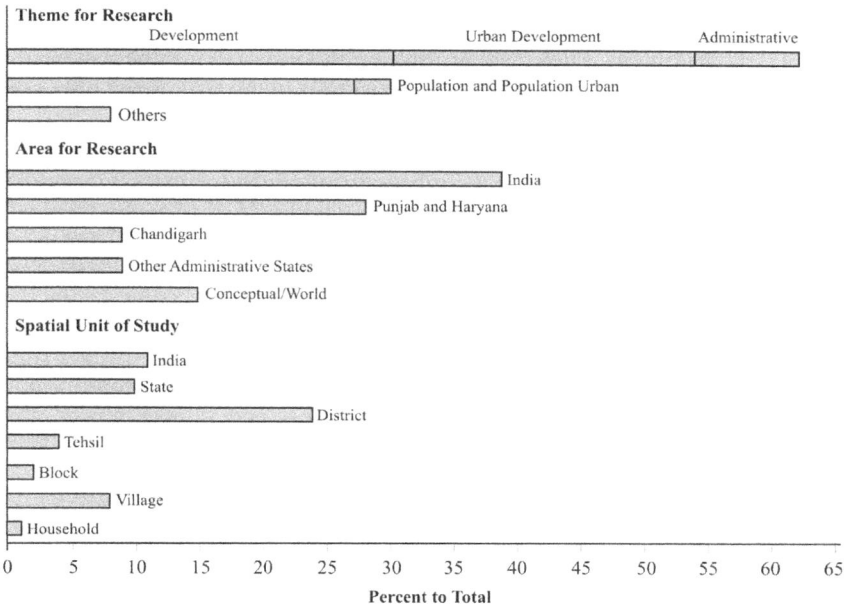

Figure 4.3 Research Terrain.

GK's work was his discipline and he built an academic landscape by adopting a heavy schedule of work. Rising in the early hours of the morning doing a lot of reading and correction, he would reach the department before anyone else and be the last to leave.

Working quietly in his spacious, sparse room overlooking a grove of eucalyptus allowed GK a head start and time to plan for the day. Parallel with official work and a stream of visitors, GK developed the knack of keeping abreast with research. Ever on the prowl for an innovative idea or an uncharted realm, so wide are his areas of interest that a censure of being easily distracted is levied on him. There are works on population growth, distribution, migration, resources, agricultural development, social development, slums and municipal services, service centres, policies, area reforms and numerous other issues. There are conceptual works, methodological expositions, reviews of literature, and presidential addresses (Figure 4.3).

GK applied his craft on various spatial scales but the nation and the area in his vicinity were his favourite. Research at the India level constituted 40 per cent of his works; Punjab and Haryana are the focus in 28 per cent of his studies, while Chandigarh, the city of GK's domicile for 50 years draws 10 per cent of GK's attention. The melange and terrain of GK's work elude a common classification. To untangle the diverse web requires patient reading where the message and not title, the approach and not data, the scope and not area are of fundamental concern. To operationalize such an abstract, self-styled criteria is not free of problems but it provided the solution. GK's works can be assorted into two broad themes: population and regional development. These two groups are not equal in number, internally homogenous or sharply bounded. The division is by no means the only one, but it reflects the evolution of GK's academic landscape.

Population geography and regional development studies arrange themselves into two temporal zones. Climbing from youth to maturity, population geography forms the first tier. After engaging GK's 15 years of work, from 1960 to the mid-1970s, population gives way to the geography of development. From maturity to wisdom, nearly three decades of GK's life are devoted to the study of regional development and this study is nourished with tributaries like urban development and administrative geography. An ecotone between population and development is that of urban geography.

If the publications are classified, if the specialization by which GK is known and what he would prefer to be known for is considered, the geography of regional development would outvote population geography. This, perhaps, is the reason why the editors of this volume zeroed in on development as the core theme, although population geography can neither be ignored nor discounted. In the trajectory of the academic scenario it is population geography that led the way to regional development.

Population geography

On 1 August 2002, GK was appointed Honorary Director, Population Research Centre at the Center for Research in Rural and Industrial Development. India has 18 such population centres. The destination of one's doctorate is rarely known but GK inherits an intellectual lineage in the field of population geography. G. Trewartha, a doyen of population geography was Gosal's doctorate supervisor; GK completed his doctorate with Gosal as his guide. This distillation of academics found apt relevance in India. Jeered as a land teaming with millions, the study of population was of critical value to India in the 1960s as it is today. GK's doctorate was the gateway to population geography.

The international boundary that partitioned India and Pakistan in 1947, ripped through not just land but people too. The partition led to issues of migration, resettlement and displacement of religious communities, occupational uncertainty and persistent trauma and pain. GK's thesis mirrors the impact of the partition on the demographic character of two border districts, Amritsar and Gurdaspur. In the lingo of geographical research GK's was a census thesis. Causal relations grounded in empirical verification and mapped spatially were the kernel of the apprenticeship for a doctorate. GK's later publications drew a cross-comparison with Orissa and Uttar Pradesh. Burgeoning growth captivated three-quarters of GK's research in population. Underlying the play of agricultural birth rate and industrial death rate of people in the country, GK states, "…no Indian state or union territory has entered the third stage of the demographic transition, which stipulates both birth and death rates below 15 per 1000." A common claim in the analysis of India's population growth is that 1921 and 1951 are years of great divide, after which the trend underwent a change in speed. GK challenged this universality. In a spatio-temporal analysis he identified that one-third of the districts of India share these two years as a watershed, of the remaining districts, 69 have 1921 and 1961, 30 have 1931 and 1971, and in 19 districts the breakpoint of growth was 1921 and 1941. GK added to the academic terrain by identifying warnings of spurts in population growth.

Concealed in the endless rows of depersonalized figures of the census volumes are the people that constitute the amorphous mass called population. GK pierced the serried rows of numbers to reach out to the less privileged group. The non-agricultural workers, illiterate females, minorities, internal migrants and infants facing mortality are focus for research. Plotting distribution and explaining concentration, GK illustrates all this by highlighting the gap between the literacy rates of rural and urban areas.

The claim of any fledging population geographer is that predicting is the kingpin of planning. Foretelling is chancy, therefore as insurance, GK employed a battery of statistical measures like ratio, extrapolation, and the growth differential methods before calculating that, "…population and class structure for Chandigarh will grow two fold from 1991 to 2020, in which the percentage share of middle class will decrease but that of the upper and lower class will rise." This research is fodder to the Chandigarh Perspective Plan and GK warns, "…if appropriate rehabilitation schemes are not undertaken, [the] slum population in the city would rise from 10 per cent in 1991 to 20 per cent in 2020." Examining the sample data for the entire state of Punjab, GK estimates that rural Punjab had over one *lakh* (100,000) unemployed and two and a half lakh underemployed persons in 1981.

That ideas evolve in fits and starts and have a situatedness in the milieu of prevailing knowledge is displayed in GK's publications. In 1965, he infers, "non agricultural workers in rural areas are strongly correlated with proximity to cities and rates of literacy." Fifteen years later, in a finer analysis of what engages non-agricultural workers he concludes, "hardly 6 per cent of the total non-agricultural workers are in services of wider regional market, 94 per cent are directly or indirectly dependent upon agriculture." The rage of the feminist movement from the 1990s in India is input in GK's reassessment of the role of women at work. He advises, "…the method of data collection by the census which needs to find ways of netting more women workers in their count, either by relating to their contribution at peak agricultural season or by assessing activities handled by women in cattle care, forestry and other such vocation." Another conviction is that, "…. all questions should be addressed directly to women" (1993). Sikhs, both inside and outside Punjab, are the only religious communities that seem to hold GK's attention until the mid-1970s. But interest widens to Muslims, Christians,

Bahais and even Nastiks in his latter publications. Popular discourse on the role and rights of minorities in Indian society from the 1990s steered GK's research to religious minorities other than the Sikhs. His most striking conclusion was that since independence the minorities have not only gained demographically but have also dispersed spatially.

GK experiments with different spatial scales. Tabulating for 2,837 villages in his doctorate, considering 300 police circles for the study in Orissa, 60 households in Salamatpur, 67 tehsils of Haryana and 72 tehsils of Punjab for demographic analysis, the district becomes his preferred choice. Most of GK's population analysis at the national level filter down to the district. Such is passion that in later years the district is transformed from a unit of study into a subject of study.

GK's adeptness with the census volumes, and the acceptability and continuity of their data, inspired 13 of his 24 Master of Philosophy candidates and two of the ten doctoral candidates to work in the area of population geography. Along with the supervised research, GK has two books, 33 research papers and three research projects exclusively on population studies, leaving no doubt that he built a substantial expertise on population geography. Yet by the mid-1970s his attention moves towards the geography of regional development. While it is difficult to determine precisely when the shift began, there are some place markers: in 1976 when the Department of Geography introduced the programme of Master of Philosophy, GK announced a special course on regional development. Confirming his mood is a quote from a paper published in 1980, "...a course in the geography of development is going to be intellectually stimulating..." Even the 1981 bulletin of information of the University of Panjab, records GK's specialization in regional development. GK's appointment at the National Institute of Urban Affairs, Delhi in 1983, carries the title Professor, Regional Planning. Thus, the seeds of the second tier of the academic landscape were sowed.

GK rarely abandons a love he nurtures, be it for people, places or themes of research. In years to come, an occasional project, lone articles or the map series prevent population studies from withering. It is a different matter that later publications on population undergo a change in complexion. Take the publication on new themes in population research and that on population geography in India. After a critical review of the subject GK draws an impressive agenda for future research. Political demography, gender analysis, population policy, the relationships between population and ecology, population and development and population and poverty are advertised as exciting topics for research. Is GK luring young geographers into the fort of population studies? Is this a list of regrets of what he could have done or is it a blueprint for himself? Who knows? After all as director, Population Research Center, he has returned to a place he left 30 years ago. In 2002 it could be a case of coming home. But at the end of the 1970s it was farewell to population and welcome to regional development.

Why did GK veer towards the geography of regional development? After all, the course of population geography that had been introduced in India by the Department of Geography, Panjab University had gained popularity. An exclusive journal, *Population Geography* was the pride of his department. All said and done, population geography offered unexplored vistas awaiting geographical inquiry. GK himself does not supply the needed insight except in a very general way. One could harbour numerous guesses. It could be that GK had practised the techniques of population and found them sterile, or it dawned on him that the problem of population cannot be divorced from the context of development, or he was looking for something new. It is difficult to trace the exact reason for his transition. New subjects interested him, but this was not just an ephemeral interest. Dislodging from population geography meant not just a shift from a theme, a solidifying specialization of his department and his revered supervisor, but it was a gradual process of interiorization. A kind of *tapas* where the inner transformation escapes our direct vision but the results are evident in the writings post-1970s.

Geography of regional development

Development has a noble place in the history of ideas. This was more so in the middle of the 20th century. Emerging from the euphoria of independence, crippling poverty, unrelenting population growth, mass illiteracy, fragmented infrastructure and a weak industrial and agricultural base confronted the young nation. Phased in slots of five years, centralized plans were envisioned as the panacea to all ills of backwardness. This thought process brought about the Planning Commission in 1950, India's apex planning body. After the first three five-year plans had run their steam, awareness for balanced regional development arose. Location specific plans for hills, desert, flood and drought prone areas; tribal, backward, and border areas; and resource regions were drawn from the mid-1970s. Regional planning held a promise for efficiency and equity as well as equitable spatial distribution in development efforts. Geographers at the Delhi School of Economics, the Indian Statistical Institute and the Planning Commission orbited the Nehruvian model of centralized national planning. GK was not among them.

Somewhat parallel, a coterie of social scientists pointed out that mere distribution or readjustment in resources would not solve disparity across space and people. Viewing regional development as a political category, in the Indian case they saw it as a legacy of colonial rule. Geographers, economists, sociologists and radical thinkers created a common nest in 1969; the Center for the Study of Regional Development, Jawahar Lal Nehru University, New Delhi. Though differing in approach, they made vibrant contributions to the intellectual underpinnings of regional development.

GK trod the path of self-learning. Trusting the wisdom of literature to provide a more than effective outline, he studied classics such as Myrdal (1957), Friedman (1964), Haggerstrand (1967), Brookfield (1975), Johnston (1977), and Smith (1977). (Google these if you have the time and the inclination.) Browsing literature and interacting personally with some scholars in the West kindled his awareness of inequality, welfare, interdependence and theories of regional development. Though GK concedes that development is to be understood as a historical process and that its future needs to be planned, for him it was not so much the past nor the future, but the present where he found moorings within the geography of regional development. GK's orientation seems to emerge from his eclectic background where he never felt it necessary to adhere to any exclusive school. He chose to see development as a cumulative enterprise where different disciplines provide useful insights for an understanding of development problems. His fascination is a coherent perspective. Even when he turned away from population, he could have moved towards research in agriculture, industry, resources or any other branch of geography. He was aware of the traps of confinement to a fragment of a discipline. Synthesis is common to the spirit of geography and development.

GK's conviction that the relevance of academics stretches beyond the portals of a university towards concerns of the state formed a core attraction for the geography of regional development. This thinking permeated through and drenched GK's subsequent teaching, research and writings. An attentive reading of his 70 odd publications and documents, numerous lectures and heaps of notes written in his neat small script make it apparent that, for GK, development was not an 11-letter word. Only one letter, "D," caught his rapt attention. The canons of GK's research rest on multidisciplinary *Data*, with *District* as the unit for analysis, where a methodology to *Devise* an index is employed to focus on *Definition*, *Disparity* and other *Dimensions* of regional *Development*. On these pegs the second tier of GK's academic landscape was clamped.

Definition

What is development? Is development about money, power, justice or freedom? Is it about acquiring goods or about being good? Can development reach all people in all places? If so, how? Such questions peppered GK's introductory lectures to students on regional development. Basic is what one should first define is his oft-repeated line. He looks at development not merely as a process but a paradigm. In geographical parlance, development denotes, "…the quality of a regional system in terms of economic efficiency, social equity, political maturity and ecological equilibrium." Further, "…a real variation in social reality can be referred to the stage of development." Collating definitions of economists, sociologists and political scientists, he states, "…the geographic concept of development is a synthesis of viewpoints of various social sciences …the routes to development are different and regionally unique, geography gives development a spatial perspective." The approach to bridge differences is evident in the debate between development and ecology. "…A golden mean between the extreme positions is represented by the concept of sustainable development." This calls for, "managing ecology for development rather than confining development for ecology". The experience of disparity marked him, never permitting him to think about homes without the homeless, food without the hungry, literate without the illiterate. To GK, the understanding of disparity is the key to any geographical inquiry in development.

Disparity

Decorated with photographs and exquisite maps, 76 articles that appeared on India, from 1947 to 1993 in the National Geographic could mesmerize any reader. But not GK. This is what he gathers, "...by coincidence or design the peripheral regions received greater coverage, focus on quiet, less populated regions of India is counterbalanced by exciting interest in metropolitan cities"(1993). Indicators are GK's mantra, "... Indicators operationalize an abstract idea, they help to understand social reality and evaluate the impact of specific programs and guide in the formulation of policies." An impressive list including food surplus to food deficit, nutritional status to status of women, number of factories to size of landholding, rural to urban and urban to rural migration, rural services to urban amenities, metropolitan cities to medium towns, infant mortality to age of marriage, joint families to nucleated ones, numbers in a rural versus urban household, among many others, are all lenses through which disparity is captured. He concludes how indicators of development are economic, while those for backwardness have a social flavour. GK's research on regional imbalance was in demand in the 1980s. Combing and combining variables to support the importance of disparity, GK cites Hartshone (1980), "...Geography essentially is an area characterizing and area differentiating science." What is the differentiation? Designing an index begins with the alphabet letter, "D," GK's spellcheck for development.

Designing an index

Aware of the pitfalls in any index, conscious that development, more so underdevelopment, cannot be couched to a mere number, the approach in GK's research is quantitative with statistics as a point of departure. To glean an objective index of disparity, GK deploys different tools. The nearest neighbour technique, the Z-score method of standardization, the mean and median centre of any distribution, residual from regression, are common apparatus. Working on an index for deprivation, homogeneity and heterogeneity and another for social, economic and ecological development, GK struggles to capture the complexities of disparity. The end product of these calculations is a number, ratio and value; while all these appear simple they camouflage the painstaking work of the gathering, processing and standardization of data.

Data

A crisp list of references is the hallmark of GK's publications. It is the foot-notes to tables and maps that give a clue to the range of data that has been mined. The Census of India, Digest of Economic Statistics, Planning Commission, National Sample Survey Organization, Centre for Monitoring Indian Economy, and documents of both the World Bank and the Asian Bank of Development occupy a favourite place. Except while indulging in a review of literature, there remains a marked preference for government reports and current policy papers over research articles and books. A geographer's work does not end but begins with the tabulation of data. A map is a final destination. GK adopts a birds-eye view to capture regional development across the sweep of India, where the worm eye roams only in a district.

District

The colossal sweep of the administrative units of India in 2001 contained 28 states and seven union territories, 593 districts, 5,564 tehsils, 638,365 villages, and 249,095,869 households. At varied times, using different data, GK has researched at each of these scales of India. Questionnaires at the household level in 42 villages form the basis of his work on employment and underemployment in Punjab; nearly 1,300 villages in the Ambala district, Haryana were chosen to study disparity in services and facilities; 965 villages in the Sirmaur district, Himachal Pradesh were the unit for identifying distribution of service centres; integrated development plans were drawn for three blocks, Makhu, Guru Har Sahai and Anandpur Sahib, while 72 tehsils in Punjab and 67 in Haryana were the unit for the study of industrial and agricultural development. His view is that micro level planning is essentially a geographer's arena, for they alone can spell location specific projects. Fetching experience at these spatial levels it is, however, with the district that GK draws close affinity. So abiding is GK's faith that from a bounded space, he questions the relationship between the characteristics of a district and its level of development. In his desire to strengthen district level planning, GK takes time to spell out a methodology for transfer of resources to the district. With Punjab as a test case he states, "...the methodology tilts in favor of the less developed districts, and therefore meets the tenets of justice and welfare to backward areas." He advocates, "...there is urgent need for intensive research on empirical and theoretical solutions to the problem of districting." (1990)

Nearly 25 per cent of GK's works select the district as a unit of analysis; of his total maps in print 60 have outlines around the district. The manageable number, comparative ease in availability of data, and the fact that the district is the basic functional unit for all administrative and development activity is the reason that governs selection. Rooted in theory, grounded in literature and verified through empirical analysis is what lends depth to the analysis of disparity. Teasing apart relationships and forging new connections GK pitches different variables in search of causal and spatial understanding. Food deficit and surplus areas when juxtaposed with type of agricultural produce and levels of urbanization reveal, "...the deficit areas are ones, which produce rice and are relatively more urbanized." (1988) When a correlation is run between level of development and cultural pluralism, "...evidently the developed states in India show a high degree of pluralism while the less developed states carry greater homogeneity in their composition" (1994). GK views a medium town as more efficient than a small town and more just than a city. But when relating occurrence of medium towns with the level of development he draws the conclusion that, "...medium towns do not come out as effective growth poles for the diffusion of the development process or for reduction of regional disparity" (1996). Thus GK draws valuable inferences and conclusions about numerous dimensions of the state of India's regional development.

GK's study of the geography of regional development is a saga that encompasses 40 years from 1961 to 1991. The absence of a collective assemblage of data, maps and analysis robs one of a dynamic spatio-temporal analysis of India. GK should have scribed a seminal text. His concern is with the map of the future, "...a new development map is on the anvil; economic decentralization, privatization, globalization, regionalisation of polity and strengthening of local bodies, will determine the contours of the future regional disparity map of India." Whatever else, GK's desire to map the 2020 regional disparities of India 2020 will manifest!

A study of regional development for GK is not an end but a means to mobilize India towards greater social and economic equity. His contribution to the city of Chandigarh and advocacy for administrative geography exemplifies a search for relevance for his discipline. It is a different matter that one gained immediate success while the second awaits success.

Success: Chandigarh

The city of Chandigarh was barely six years old and GK 15 years when the two met. One among a 120 planned cities of post-independent India, the blueprint for Chandigarh was drawn in 1949. GK had an opportunity to see the sights of this city for the first time in 1955. Returning with a prize at a declamation contest held at Gandhi Memorial College, Ambala, and Hindu High School, Khaitol, Chandigarh was an en route destination. Halting for a brief three days, as GK had to return to his school and home in Shimla, he recalls, "...there hung an air of promise on the recently etched dust laden roads, the plinth of large structures awaiting construction, it was a feast to behold Chandigarh, at the doorstep of the temple of the goddess *Chandi,* move towards its journey to completion." The design of Le Corbusier and the gigantic experiment in architecture and town planning had obviously impressed the teenager. The day was to come when as tribute GK was to pay his compliments to the city. The day came.

The city was 50 years of age and GK 59 years old. To commemorate the Golden Jubilee of Chandigarh, the International Union of Architects, Paris held their annual conference there on 9 January 1999. The chief guest was President K.R. Narayanan of India. The offering was a bouquet of 65 handpicked and self-designed maps held together in a rectangle-shaped wrapping called, "Inner Spaces-Outer Spaces of a Planned City: A Thematic Atlas of Chandigarh." The sponsors: The Administration of Chandigarh. The author: Gopal Krishan.

The event was a milestone, a celebration not only on age but the performance of the city. Indian cities are a mixed bag, the good laced with the bad. In a survey of The Best Cities to Live published by *Business Outlook* (28 April 2003), Chandigarh topped the charts. On three counts, job and income, financial infrastructure and consumption, it stood first. Contributing in ways both big and small, it is the people of a city who make the city. A few make meaningful contributions, most are critical, while the majority live in apathy. GK is a citizen of the first order. As a geographer he has done more than his bit for the city.

A permanent address for over 40 years, Chandigarh bewitched GK at the very beginning and penultimate phase of his academic life. It was in 1963 that GK first ventured to study the relationship of Chandigarh with its rural hinterland. Research on the umland captured by a planned city was published in 1968. After a gap of another five years, his third piece of research on the commercial plan of the city came in 1973. Once again as a team, he contributed towards the compilation of the Gazetteer on the Union Territory in 1976. Thereafter for over a decade GK forgets his hometown. The population projection for 2020, an important ingredient into the perspective plan of the city is a work of 1994. The atlas was completed in 1999. Comments on the periphery of the city, reconstruction of its municipal wards, and a prognosis of the emerging scenario all appear after 2000. In a number count GK has an estimated dozen publications on Chandigarh. Only one of the 24 Master of Philosophy Students he supervised and one of his ten doctorates selected the union territory as a unit for study. It is ironical that where he has published the least he is known the most. The nature of publications matters more than sheer numbers.

The current condition of Chandigarh, once an ambitiously planned city would inevitably lead to indignation on the part of someone associated with urban development. Rather than become embittered, GK chooses a multipronged agenda to bring both the city people and the administration out of limbo. Sensitizing students with field visits across the city, creating an informed public through writings in newspapers like *The Tribune, Indian Express* and *Dainik Bhaskar* and delivering lectures at different podiums are sundry offerings. In the final it is his contribution to the planned development that epitomizes GK's success. Chandigarh is a microcosm which allowed GK to see things as a whole, although on a small scale he tried out many forms of activity. The projections about the growth of population, the survey on slums, the perception of city dwellers, a critique on the policy of the periphery, and the atlas all levitated him to the circles of city administration. Nominated by the president of India, as member of both the Punjab and Haryana State Planning boards and the steering committee of the Chandigarh Administration gave GK a handle to push the case of geography and its role in urban and regional development. Today "city beautiful," as the given nickname for Chandigarh goes, the Perspective Plan for 2020 envisioned that it would earn the subtitle "city of excellence." As a harbinger to excel, GK's role sustains. Research on Chandigarh and work with administration is an accolade of success, but GK's persuasion about the importance of administrative geography as a discipline worth study in India, awaits success.

Awaiting success: administrative geography

Similar to a lawyer in court, GK makes a strong appeal for the importance of administrative geography as a separate branch of the discipline of geography in India. The crux of his argument, "...to facilitate a development process, India needs to have a spatial arrangement of administrative areas where an unwieldy size or irregular shape or a truncated resource base of a unit does not hinder the smooth operation of development activities and cause waste in already limited resources." He quotes, "... any regional account begins with an administrative map of the study area, most of the research in human geography is based on data by administrative units and every geographic analysis recognizes administration as a crucial explanatory variable" (1990). But there are few takers. He toes another line of defence, "...my basic proposition today is that the essential virtues of efficiency, equity and effectiveness in administration can be best achieved with the help of cartography. Administrative efficiency depends upon an appreciation of ground realities for which [a] map is a potent tool. Historically maps were used for war, the contemporary world of administrators have to contemplate their use for peace and development" (1997). Even though he appeals to the reason of his audience rather than their emotions, his petition fails. Some argue administrative geography is an extended limb of political geography, others hold the view that administrative concerns are outside the purview of academics, while for some its theoretical underpinnings are difficult to digest at the outset.

Perhaps the empiricist court needs proof. GK makes an attempt: he calculates the locational eccentricity of state capitals (1976), highlights the issue of territorial jurisdiction of towns in Punjab (1983), researches urban–rural relations and states that every administrative unit should cover one core city or town which is functionally integrated with its surrounding territory. He suggests a set of criteria to determine the optimal size of a district and puts it to test in Punjab, "...a district should approximately be 2000 square kilometer area and the state should have 25 districts"(1992). To add muscle to his argument GK recasts a new map for the Chandigarh Municipal

Corporation by stating that electoral wards should be increased from 20 to 40. These should be constituted on single sector basis – where the rule of thumb is that a sector with a voting strength of around 10,000 could automatically be raised to the status of a ward (2002). Still there are few followers. In spite of the fact that GK's research on administrative geography is published in journals like *Geographical Journal*, *Economic and Political Weekly* and *Asian Profile* and despite the fact that as president of the Indian National Cartographic Association in 1997 he gets a chair to air a conference on Cartography and Administration; that two of his ten doctorate students have published independently on this theme; his atlas on Chandigarh is a creative work sponsored by the Chandigarh Administration; he draws experience from a decade of sustained association with the Indian Institute of Public Administration, Delhi; has authored a monograph under the eye of strict supervision at University of Cambridge; and that since economic reforms, constitutional reforms and political reforms have made way in India, administrative area reforms are likely to follow heel, the torch GK lit for administrative geography has not been taken on by others.

GK is rarely out of touch with his own reality. In a status paper for the International Geographical Union he acknowledges, "…Administrative geography is in a nascent state in India" (2000). Empathizing with "his" theme a bit perplexed he writes, "…this non-event is difficult to understand." (2000)

Perhaps it is not so difficult to understand. If only GK would accept that diffusion of any innovation takes its fair share of time, unlike themes which find a ready market and cut a quick swathe. He should also not overlook that a booster to a course is the ability to push it within a national curriculum such as that of the competitive exam. Perhaps what he needs to remember is that a subject is likely to fall low if one does not till it with vigour.

Experimental and progressive, GK's academic landscape carries its highs and lows. Creditable is the unsnapped vigour which contours the retinue of his research. Immeasurable in size, bonding all the way, keen to excel, GK is the architect of his own pace and theme of work. What could be an appropriate nomenclature for such a map?

Name for the map

A search for a title that could sum up all that *this* map signifies takes one through a list of impressions, some phrases, idioms and homophones. Naming *this* map was not easy. What is surprising is that for christening a newborn there are half a dozen books in the superstores, but for an appropriate name for a lifetime's work there is no help book! Here are some offerings from the search engine of my mind: GK – Life and Works, Vision and Mission, Busy and Dizzy, Philosophy and Research, Trial of a Geographer, Inner and Outer Spaces, Geographer with a Difference, The Development of a Geographer and the Geography of Development, Ways of an Indian

Geographer, From Research to Relevance, Prolific versus Profound, Karma and Dharma of an Indian Geographer, Good in Pursuit of the Best, In Search of the Self, Contours of Life, A Biographical Reflection …thus went the list, on and on and on. None captured the essence or seemed appropriate, so all were junked to the recycle bin. It is not that GK is too complex or too diverse to square but the fact is that he is the way none other is. A substance and style so much of his own, the only fit that seems to fit is *Map: Gopal Krishan.* At times what is most visible is that which is difficult to see.

All who have interacted and known GK would agree that he has distinct ways. Take his conversation; rarely extending a few pithy sentences tinted with a hue of hesitancy, words like grand, deserve, excel, perusal, perennial, integrity, and creativity, are gems he patronizes. Said with heartfelt concern, despite being overused, their message endures. This is vintage GK.

To research

Questions form the core of GK's style of research. If ever on a dead end for a relevant theme for research, GK's publications are a gene pool of over 300 questions. Nearly every publication begins and ends with a question. The enchantment with questions runs into even the main title. How does India stay integrated? Is India over-urbanized? Question of regional identity? Has the Chandigarh Periphery become a peripheral issue? Reflecting curiosity, questions range from the abstract to the most pragmatic. The axis of most questions tilts towards problems of the geography of development and with years, the questions outnumber answers, confirming that the more one learns the less one knows!

GK questions established myths and has the propensity to swim against the current. Such a mindset has been evident since his younger days. In a publication authored when he was 23 years he mentions with all certainty, "…all winds which undergo direction reversal with season are not monsoons … naïve concepts are no longer tenable" (1963). He notices the paradox of Punjab which, "…despite being the highest per capita income state has the largest immigrants." He adds, "…an inland by location, Punjab behaves like a coastal state" (1997). To muscle financial and political benefits the government of Punjab is keen for recognition as an industrially backward state. Unwilling to step in line GK's research concludes, "…there is scope for further industrial development, but in no way can Punjab be declared as an industrially backward State." In a similar vein he does not kowtow to the municipal administration and blames, "…the failure of the basic service provision in cities is a glaring failure of the public sector management of the services." With respect to the augmentation of municipal services and the role of privatization GK's stand is, "… services are rendered and not sold."

The power to question and view a situation upside down is GK's research style. In a policy paper on the slums of Siliguri he observes, "…it is generally viewed that a slum is a problem, but slum dwellers perceive the slum as a solution." Analyzing population growth he affirms, "...the demographic transition model does apply to the Indian situation but the western quantitative parameters do not apply." Demythifying India's supposedly grim population reality GK states, "...Rapid population growth has not arrested economic growth but slow economic growth has caused rapid population growth." On the bedrock of data GK shatters myths that it is the urban population that adopts family planning not the rural population, and Muslims are indifferent to family planning. GK chooses to systematically prove that, "...the heartland periphery and coast inland contrasts are more striking than a north-south divide." In support of his drive to explode cherished beliefs he affirms, "...Some of our deep imbedded myths have to be given up .... Only correct perceptions are capable of yielding the right kind of policies." Altering cognition is a fundamental step towards a process of development.

While questions are starters and terminators, GK grounds the arguments of research either within a national policy or the axioms of a philosophy. The Integrated Development of Small and Medium Towns launched by the government of India in 1979 is the context in which GK analyses the catalytic role of these towns in the countryside of Punjab, Himachal Pradesh and India. He observes that Punjab pre-empted this programme. Tracing the concept of the national capital region alongside a perusal of the Draft Regional Plan of Delhi GK adds, "... Delhi is bound to grow with or without a plan for the National Capital Region. The only solution is to lay a transport network to expand the city in different cardinal directions." The success story of local area development in Sukhomajri, a village 30 km to the north-east of Chandigarh is well known. However what is lesser known is the ephemeral nature of its success. Evaluating the failure of Sukhomajri, GK not only conducts personal field trips but also roots his analysis in the philosophy of a cooperative and concludes, "...Sukhomajri did not graduate into comprehensive development, it was bound to fail." Testing Christaller's Central Place theory in the service centres of the district of Sirmaur, Himachal Pradesh, GK proves that, "...several parameters of Christaller's model which are essentially pertinent to an isotropic surface, are applicable even to a mountainous region."

Contextualisation and a profound documentary understanding of facts are at their best where GK analyses spatial disparity across India. Flushing the argument within a political and economic framework, referring to central and state policies, citing the promise and failure of the five-year plans GK sums, "...Striking regional disparities stand out on the development map of India." It is a strange situation where the geographical heartland or core is weak and its shell or periphery is strong. India has a relatively more developed west and less developed east. Its north-south characterization is more along cultural than economic lines. He analysed this spatial pattern against the grain of wealth of India's physical resources. Practised with care GK displays subtle mastery at knitting macro with micro.

Toying with ideas, raising questions or applying a model or critique is not the main aim of GK's research. By transcending the role of knowledge from information to education to wisdom, GK's research seeks to direct itself to formulating policy to correct the socio-economic and spatial imbalances. To spell a policy, GK fuses statistical data with primary fieldwork. His astute observations use the casual remark of a farmer, housewife or agricultural labourer to obtain insights that no quantitative research would reveal. He integrates knowledge so acquired to draw generalizations. Future geographers would prize GK's works for the honesty with which he went all out to make geography relevant.

GK's generation of geographers were buoyed along a great wave of studies of disparity and balanced regional development. But within its linearity, GK's research carries the element of progress. As thinking shifted from centralized towards decentralized development GK was quick to review the 73rd and 74th Amendment of the Constitution. Positioned between centralization versus decentralization, role of government versus non-government organizations, public versus private and global versus local, GK affirms, "...plan accomplishments have fallen short of expectations, today there is a streak of irony, while the new economic policy may centralize development at more advantageous location one hopes the political decentralization will counter such development" (1999). A critique levelled at some of GK's publications is that they are wide rather than deep. GK explains this as the intellectual parameters of the discipline of geography. Addressing the forum of the National Association of Geographers, India at Dharwar, his presidential speech resounds, "...Geography as a discipline is obliged to be all inclusive hence less deep .... It is not that rigorous in training hence less specialized and it is polite by temperament, hence less active" (2000). While GK's style of research is loaded with questions, incisive and polite, his style to administer is to lead.

To lead

The date of 31 March 1987 was a historic moment in the life of the Department of Geography, Panjab University. On the evening of this day Gosal retired and before noon of the next GK became chairperson. The circle of time hands the baton to the next generation. Inheritance, it is said, is often squandered by the next generation, some maintain the status quo, few lead ahead.

As chairperson GK wanted the Department of Geography, Panjab University to be India's premier institute on the map of the discipline. Competition was tough. While departments such as those at Aligarh, Patna and Benares had the advantage of age, some, like the Delhi School of Economics, the Centre for the Study of Regional Development at Jawahar Lal Nehru University, the University of Bombay and Calcutta University had a metropolitan attraction to retain talent. The Department of Geography, Panjab University had earned its share of merit. Association with organizations like

Census of India and Ministry of Defence and the status of Departmental Research Support from the University Grants Commission were its bonus. Yet it needed to blossom out of its regional and inward-looking shell. There were other troubles. Compared to the 1960s, slack academic environs began to erode the work culture in universities from the mid-1970s onwards. The national entrance test and merit promotion scheme injected apathy towards scholarship and universities were in a mode of decline. Notwithstanding such a scenario, a prize that came to the Department of Geography was the Special Assistance Program; a scheme by the University Grants Commission to flush funds into a few select departments that showed potential for growth. Only four out of 78 departments of geography in India in1990s were recognized as special. The programme was perfectly timed for GK. On the very next day his office of chairperson drew to an end he stepped in as Coordinator of the Special Assistance Program. An event similar to the day he walked out of the class to the classroom.

Funds of around five million were funnelled into the department making the programme truly special! A coordinator can continue for extended terms whereas a change of guard every three years is compulsory for even the best of chairpersons. While GK was a passionate advocate for the system of rotational chairpersons, he still held the office of coordinator for 12 long years and why not? The programme bore fruition of GK's vision to carry the department towards becoming a centre of excellence.

Money, we will all agree, does not run a programme. Managing money at best will upgrade and refurbish the department with state-of-the-art machines, some odd infrastructure and allow travel grants. GK's agenda was not limited. His intent was to convert the department into an intellectually vibrant niche. Organizing annual seminars on contemporary issues and inviting select audiences paved the way for healthy discourse; keeping crowds at bay and laying more emphasis on the quality of deliverance than attendance by rote. I am probably the exception to dare to gatecrash. News that the Department of Geography, Panjab University was good at organising seminars had reached the Delhi School of Economics where I was lecturer and the brochure of the Seminar on Geography and Public Policy to be held on 7–8 March 1996, at Chandigarh caught my attention. I had read and heard about GK, but never met him or spoken to him. Plucking up courage I called him a day before the seminar. In his usual mild style he granted permission, but sugared the reality that I would be an unofficial entry, without allowances for travel. I am not GK's student nor his colleague; we share no joint authorship or claims in a project. Yet the fact that destiny chose me for writing this is proof that I owe allegiance. Seminars at the Department of Geography, Panjab University are channels for growth. Held annually, while the three-day events occupy the department for months of pre-seminar preparations, what is active upfront is the array of guest faculty.

Handpicked eminent scholars across the country visited the department during the 1990s. Setting a style of scholarly interaction and gentle hospitality, GK strengthened the activity by injecting an international element into the programme. The log in the visitor's handbook of the department lists 63 guests from 1992 to 2002. This is around six per year. If the academically active term is considered, then every month the department was host to one scholar. Most are residents for a week. Thus, research students and faculty drew benefit from a scholastic guest for 40 days in a year. The rich hinterland of intellectual resources and distinguished stream of visitors catapulted the Department of Geography, Panjab University to a leading position. When GK handed in the baton of his 12-year tenure as coordinator, Special Assistance Program to the next in line, modern equipment, an extension to the main building, a course on geographic information systems, and a high reputation were its bank balance.

Some say that GK could have been more democratic to enrol a larger team within his department. In a SWOT analysis of himself he concedes, "… I do try to involve the entire team but soon ignore stake holders who show no interest, reaching the ultimate goal and moving towards a sure success is my prime aim." GK's ability to remain outside the tangles of obtrusive arguments, away from cliques and the politics of inclusion and exclusion helped him win more allies. The Panjab University has 52 postgraduate departments. The compliment, "…department of geography is the best run on the Campus" by the vice chancellor of the university on 12 February 2001, the inaugural function of the seminar of the department, totes up another score in the list of arenas where GK excels.

The Panjab University is designed as a residential campus, interspersed between spacious lawns, teaching and administrative blocks are the hostels. Writes the Dean of Students Welfare, "…of the 10 hostels on campus, the boys hostel called A.C. Joshi Hall is the best run." The letter is dated 1980, the year GK was warden of 300 resident students in this Hall for five continuous years.

Inspiring groups of different size, composition and technical expertise GK successfully completed 20 research projects. Nowhere is the team spirit more inherent than in the atlas project. Conceptualizing and publishing the atlas within six months was no mean feat. Including data collection, field visits, editing and printing this is a rate of output of two maps per week. While GK threw himself full throttle into the work he admits, "…the cartographic staff and research students were a strong scaffolding." My digital recorder captures the feel of one among the many in the team, "GK likes to give and take the best. In those six months, I worked not for but with him. I owned the Atlas." It is in a bed of fertile support that ideas flourish and flower. If GK accomplished his atlas in six months, he led my book to its destination in 180 days. Leaders may be born but leadership is cultivated. In an increasingly atomized society, GK enrolled many among his colleagues, students and others as co-pilots. Conquering all through prodigious work, eloquent scholarship, and an uncompromising set of values GK's single goal was to take his discipline and thus his department towards the heights of excellence. Managers are people who do things right, leaders are people who do the right thing. Most departments of geography in India are mismanaged; some are at best, managed; few are led.

On the dawn of 31 July 1960, GK climbed 44 steps to the second floor where the Department of Geography is located. He was 20 years old, a student carrying a freshness of youth and a dream of the life ahead.

At dusk on 31 July 2002, the Department of Geography bids him farewell. Amidst a row of candles that lit these very 44 steps GK bids adieu. He is 62 years of age, carrying a face that is much weathered and an island that swallowed many a storm.

Yet etched in the crow's feet around his eyes and the laugh lines along the smile is an incorrigible optimism and a deep satisfaction with a life well spent. A lifetime worth a map.

As a geographer in India GK led a department, committed to a discipline and worked for his profession. If a few more could follow his way, the shape of the discipline in India would acquire a new map.

Acknowledgements

Mapping, especially a difficult terrain, requires teams and tools. My students Anuradha, Roshani, Deeptima, Punam, Meeta and Neeti helped compile and catalogue his works. GK's family were most gracious in sharing anecdotes and perceptions.

5   A master geographer–planner
K.V. Sundaram

There is an art of living. There is an art of doing. There is an art of being. K.V. Sundaram mastered the art of living, doing and being a geographer–planner in India. Sundaram tasted the waters of planning within the portals of an academic environment of teaching and research and tested its acceptability within training programmes and amidst interdisciplinary teams. Planning for development at the national, regional and state level, his experience embraces town and country, rural and urban. Contributing key ideas Sundaram served on the boards of the Town and Country Planning Organization, Planning Commission and State Planning Boards. Premier organizations such as the National Institute of Rural Development, Hyderabad, the Lal Bahadur Shastri Academy of Administration, Mussoorie and the Indian Institute of Public Administration, Delhi sent him frequent invitations. Sundaram's contribution to spatial policies earned him a place on scores of local, national and international committees. His name appears amidst the architects of regional plans like that of the South East Resource Region, Dandakaranya, Bastar and the Rajasthan Canal Project as well as those who designed the first Master Plan of Delhi. Being on the payroll of the United Nations Organization, he lent consultancy to numerous international bodies. His advocacy of decentralized planning took him to serve and advise not just the neighbouring countries of Nepal and Bangladesh but also countries like Uganda and Ghana in Africa and the Philippines and Malaysia in the Southeast Asian region.

Sundaram's kitty as a geographer–planner is brimming with nearly 300 publications. Decorating his office are innumerable honours, felicitations and awards. Starting out with a basic degree in geography, Sundaram scaled key posts and sailed smoothly through many rough seas. What were the circumstances that cajoled him into the profession of a geographer–planner? What were Sundaram's merits that allowed him to cut ice in this discipline in India? What are the benchmark contributions made by Sundaram to this field of knowledge and practice? What was the milieu that built the foundation on which he led a purposeful life? A life that does not just take the world as it is, but has a desire to leave it a shade better. How does one master the art of a geographer–planner? Is it Sundaram's DNA? Is it the role of chance or that of choice? Is it the time-space context or is it his style of work? Many factors need to be unveiled to gain even a glimpse of truth. Inheritance is usually a good place to begin.

Inheritance

Sundaram was born in the early hours of Saturday 30 November 1929. The year sticks in memory. One major historical event that occurred in 1929 was the crash of the stock market which plunged the world into a deep financial depression. But for India it was a year of hope. Gandhi met Jinnah and Patel at the Sabarmati ashram on the outskirts of Ahmedabad, Gujarat, to discuss India's plans for freedom.

Just as the date of Sundaram's birth is easy to remember, his birthplace is also a marker on the map. Sundaram was born in Calicut. Flanked by the Arabian Sea in the west and the Western Ghats in the east, Calicut too finds a place in world history as the port of disembarkation for Vasco da Gama, leader of a trade mission from Portugal. Therefore the meeting points of the discovery of the first all water trade route between Europe and India on 20 May 1498, was Calicut. The discovery of this sea route by the Portuguese explorer, Vasco da Gama, paved the way for British rule in India and thus changed the political destiny of India.

In the year of Sundaram's birth, Calicut was part of the Madras Presidency, as the state of Kerala was only formed in 1956. But today if one were to locate where Sundaram was born one would have to look for a different place name. This is because in 1993, English names in the districts of Kerala were modified to reflect their pronunciation in Malayalam. Calicut thus became Kozhikode. Situated 900 km to the south of Mumbai, Kozhikode is the third largest city in Kerala, the southernmost state of India.

If history loads the date and place of Sundaram's birth, his name carries the hallmarks of geography. Fortunately, a name in India is clue to a lot about the ancestry. The surname "Sundaram" belongs to the Tamil Brahmins. But castes have subtypes. Sundaram belongs to the Palakkad Brahmins whose name is lent to the Palakkad district of the state of Kerala. Palakkad lies near the Palghat gap, which is a 30 km long natural depression bifurcating the folds of the Western Ghat. It connects the state of Kerala to that of Tamil Nadu in the east. It is said that centuries ago, at the invitation of Palakkad's ruler, the Tamil Brahmins migrated from Thanjavur in Tamil Nadu to Kerala and settled in 96 villages called *agraharams* in different parts of the state. They were basically priests. Sundaram belongs to the numerically powerful subgroup of Palakkad Brahmins called *Ashtasahasra* who came from a village called Ennayiram in Tamil Nadu. Incidentally Ennayiram was a flourishing educational centre during the 9th to the 13th century.

It is hard to reconstruct the journey of our ancestors but the letter "k" that appears as a forename to Sundaram's name provides some evidence. It is customary for the second son of a Tamil Brahmin to carry the name of the grandfather. Sundaram's grandfather belonged to the Kavasseri caste.

From this ethnic and caste group, Sundaram inherits his wheatish complexion, frail 5′4″ height, fluency in Tamil language, vegetarianism and a taste for South Indian food. In terms of intellectual acumen and social status Brahmins claim the top-tier of the four castes in India. Sundaram attributes his characteristics of aloof pride and detachment to his parent's caste and upbringing. But perhaps more than date, place and caste, what stands out in Sundaram's case is the ambience of learning into which he was born and raised.

Sundaram's great-grandfather, Vaidyanathan, born in 1852, was a *jyotish* and a fluent orator. His grandfather, Kavasseri Appadurai born in 1877 was a graduate in education and surveyor with the government in Kavasseri. Sundaram's father K.A. Vanchi Iyer (1901–1984) was a graduate in physics from a foremost institution of India, St. Josephs College, Tiruchirappalli. When the college celebrated its 160th anniversary in 2003 no other than Dr A.P.J. Kalam, the then president of India inaugurated the celebrations. And why not? Kalam was an alumnus of St. Joseph! Moreover Sundaram's father was not only a renowned physicist and a stern task master but also the headmaster of the school in which Sundaram gained formal education. In 1901, the year Sundaram's father was born, India's male literacy was 5.53 per cent and a generation earlier it would have been in the decimals (Census of India 1901). With these as a parallel, Sundaram's was a family that was clearly ahead of the times in the arena of education in India.

Education was viewed as a privileged means of personal empowerment. Drummed into Sundaram from early childhood was the importance of being well versed coupled with hard work. Sundaram recalls that his education was marked by an absence of school friends and out-of-school visits were unknown. Perhaps the motto "simple living and high thinking" best fits the Sundaram *parivar*. That a high premium was placed on education is evident from the fact that among the five siblings, Sundaram's two brothers are doctorates, two are scientists and the only sister is a matriculate. The legacy continues. Sundaram has six grandchildren, two of whom are enrolled in a doctorate and others are pursuing medical, finance and engineering studies at leading universities in the world. If to learn was a family inheritance it was the association and contact with the Chettiars that governed Sundaram's path to formal education and later his engagement with research and teaching.

Formal education

The Chettiars

The Chettiars are world famous for their recipe of the spicy Chettinad chicken. But there is much more to them than gourmet food. Chettiar, means people who collect and donate. They are a trading community. With the opening of the Suez Canal in 1869 the Chettiars of Tamil Nadu saw a great opportunity in the trade of rice from Burma (now Myanmar) and Ceylon (now Sri Lanka). Many flocked to Rangoon, the main port on the river Irrawaddy. Here they acquired great wealth. But in spite of the material prosperity, the Chettiars invested a large part of their finances in setting up schools and education institutions in Devakottai, the heartland of the Chettinad – many mistake it to be Karaikudi; the latter is no doubt bigger and more prosperous, but Devakottai has the largest Chettiar population. Even today, when compared with the national average of 59.5 per cent, Devakottai's average literacy rate is 80 per cent.

In 1934 Sundaram's father was invited to be the headmaster of Nagarathar Sri Meenakshi Vidhya Paripalana Sangam Higher Secondary School funded by the Chettiars. The family thus moved from Calicut to Devakottai – incidentally, the same year that Gandhi laid the foundation stone for a *harijan* school at Devakottai. Sundaram was eight years old. Although surrounded by the wealthy, Sundaram saw that his father commanded great respect. This reinforced the virtue in learning. By the time Sundaram had completed his high school education he had acquired the taste for education, discipline and the need to focus.

In 1942 when Sundaram was nearing the completion of high school, the All India Congress Committee passed the Quit India resolution. Sundaram was far too young to participate in or engage with the freedom movement of the country. Until the mid-1950s Devakottai did not have an institution for intermediate education – for those unfamiliar, intermediate is a two-year degree acquired after the completion of 11 years of schooling. Sundaram was 16 years of age and the father was reluctant to send his lad too far for the intermediate degree.

The Rajah College is located in Pudukkottai, a short 54 km north of Devakottai. It offered two streams of study: group "A" consisting of mathematics, physics and chemistry; while group "B" comprised mathematics, logic and history. Following in his father's tracks Sundaram offered for the sciences and selected the A group. Those were years when neither geography nor planning was offered as a course of study. While Sundaram received the choice of stream he wanted to pursue, he suffered frequent bouts of illness at the Rajah College. The college is located near a river called Thamir Bharani, the water of which was used for drinking purpose. The name of the river certifies that the waters carry a high content of *tamba* or copper. Perhaps its prevalent quality did not suit Sundaram. Therefore for his second year of intermediate studies Sundaram was sent to the Madurai Diraviyam Thayumanavar Hindu College, Tirunelveli.

By the time Sundaram completed intermediate, India had gained independence. In 1948, when Sundaram was about to join a graduate programme of geography, Mahatma Gandhi was shot while going to the prayer meeting. The headlines of every national and regional daily carried the message of shock and grief. Little did Sundaram know that in years to come he would be a staunch supporter of the Gandhian model of development. The impact of independence was felt by each and every place and person in India. But each and every place and person did not experience the same impact. Strange as it may seem, even Sundaram's shift from physics towards geography as the subject of study was influenced by the promise of a free nation.

The promise of geography

Sundaram had never studied geography at school, held no fascination for maps. He was not even inclined to travel and visit places. Yet, in 1948 he found himself sitting in the University of Madras amidst the first batch of to be geographers. It was a reluctant choice. If Sundaram had his way, he would have followed in his father's footsteps and studied physics. While some routes are difficult to explain, in this case the hope that geography would build a promising career was a sure attraction. After all as for any young boy in a professional family the justification for choosing a subject in college is that it leads to a good job.

The fervour of independence unleashed the need to introduce new subjects in universities which would help towards building the nation. The British had left a rich legacy of institutions that were engaged with geographical themes. The Survey of India, Geological Survey of India, Department of Meteorology, Census of India, and the Anthropological Survey of India are examples of centres that were surveying the land and people and producing maps, gazetteers and other documents. Geography was thus perceived as a subject with a promise. While there was a felt need for the discipline, "geography in the 1950s had a nebulous presence…the total strength of students pursuing studies in higher degree classes is approximately 1000, with a total teaching staff of 74" (Stamp, 1946 in Kapur, 2002a). The next two decades saw steep growth. The departments offering geography grew in number from 17 in 1950 to 48 in the 1970. Riding this tide, the Department of Geography at the University of Madras which had begun with a diploma course in 1932 declared 1948 as the year for geography. In the same year it offered a graduate course in geography. The department thus set out to collect a class who would be willing to study this subject. Under the intercollegiate cooperative scheme, 12 students were admitted: four each from Queen Mary's College, Presidency College and Vivekananda College to take a course in geography at the department. Sundaram was a candidate from the Vivekananda College.

Adding shine to the promise was the faculty that came from abroad. The start of World War II (1939), the bombing of Pearl Harbor (1941) and Rangoon (1942) generated widespread fear in the regions surrounding the Bay of Bengal. The University of Rangoon witnessed an exodus of Indians who were teaching in various departments. Searching for a congenial habitat three among them were absorbed to build the geography department of Madras. These were George Kurian, the physical geographer and later the founder of the Indian Geographical Journal, Dr R.S. Mani the oceanographer, and Dr (Mrs) Irawathi, an anthropologist who taught human geography.

Sundaram recalls that these geography teachers were an exceptionally dedicated lot. It was in their enlightened classes that Sundaram was introduced to the texts of D. Stamp, A.E. Smailes, J. Bruhnes, J. Mumford and A.K. Lobeck. Since India had been a British colony, geographers from the UK were more popular.

Coincident or not, the year Sundaram graduated, Dr V.L.S. Prakasa Rao – the first doctorate in geography from an Indian university – joined the University of Madras as a reader. Prakasa Rao's enthusiasm for research was infectious and multiplied manifold Sundaram's appetite for a doctorate. Therefore, while in the near epidemic of college openings all in the batch of 12 were absorbed as lecturers, Sundaram preferred to enrol himself as a doctorate candidate under Rao's supervision. After all, as Sundaram recalls a research stipend of Rs. 60 was a good enough income even if a lecturer's position drew in Rs. 150 per month. While the philosophy of home has a role, the pragmatism of the world also has allure.

Again the Chettiars stepped in. This time the contract was to seek not the father but the son. The call was from none other than the famous Alagappa. In India, scholars, great personalities and achievers are honoured by issuing commemorative stamps. Such was the fame of the Alagappa Chettiar that on the 6 April 2007, the Chief Minister of the state of Tamil Nadu released a stamp paying homage to him. How and why did Alagappa seek out Sundaram? What is the nature of contribution that both made to geography?

The Alagappa

When celebrating the theosophist, Dr Anne Besant's centenary, the vice chancellor of the University of Madras urged industrialists to start colleges for educating India. Alagappa, a wealthy Chettiar, heeded the call and within three days donated 300 acres (1.2 km²) of land and 1.5 million rupees to set up the Alagappa Arts College in Karaikudi. The winds that blew in favour of geography awakened in Alagappa a desire to establish a department at his college. But for this he needed faculty. The obvious window for inquiry was the University of Madras. As luck would have it, by the time Alagappa contacted the department, Sundaram was the lone geographer who had not sought employment. Therefore, Kurian, chairman of the department, put Alagappa in touch with Sundaram.

Alagappa was no ordinary man. Sensing Sundaram's ardour for research, along with the roster of teaching Alagappa offered all support to Sundaram to pursue his doctorate. The proposal could not be refused.

In a way Sundaram's fervour to build repute for geography matched Alagappa's vision of education for India. When two minds of great hearts meet, the sky is the limit. The world was soaring great heights. On 29 May 1953, Edmund Hillary and Tenzing Norgay had scaled Mount Everest. As the world applauded, for Sundaram this event became an opportune time to popularize geography. Sundaram wanted to advertise that climbing the highest point of the earth's crust measuring 8,848 m was a geographical feat. The method he used was novel. When Sundaram taught the class of physical geography to his batch of students he often used three dimensional models of the landscape. Trying to build a model of Mount Everest was well worth a try. Whereas the students were greatly enthused it was Vedappam, a lad, less a lab assistant and more a sculptor in the department who helped clinch the project. Experience had taught Vedappam that mud collected from a snake pit is soft and more malleable to cast models. As a team, Sundaram, Vedappam and the students, fashioned a large 3D model of Mt. Everest. Such was its awe that to give it wide publicity Alagappa planned an exhibition in the college. The model was unique and attracted visitors from surrounding villages and towns. After all in days when television was unknown, and laptops and the Internet were unheard of a virtual reality model did hold fascination! So successful was the two day exhibition held in January 1955 that an eight day repeat was organized in November of the same year.

A slender booklet in Sundaram's filing cabinet called *The Exhibition* documents the laudatory observations. Some are worth reproducing: "I was Mr. Alice in the veritable wonderland created and maintained by the students of geography"; "I have never seen a geography exhibition before. But after seeing this, I could imagine that geography is a wonderful subject to be studied"; "I went around all the exhibition rooms of the geography section …how I wish that I were also a student of this geography department and shared with these students, the almost infectious enthusiasm transmitted through their lecturer."

Perhaps it is this fascination for models and the success of the exhibition which sowed in Sundaram the seeds for a museum of earth care for India in the later years of his life.

In the meanwhile, as the years of teaching flowed, Sundaram spent many evenings burrowing and browsing through the stacks of geography books. Whereas the papers he taught were elementary it was a time that allowed him to crease out much that was lacking in terms of becoming well versed with geography. Frequent visits to Prakasa Rao, his doctorate supervisor, were forays into advanced geography and maps. Prakasa Rao emphasized that one good way to grasp spatial interactions was to put it on a map. Under his tutelage thinking about the discipline and practice of geography was gained. Along with teaching and research Sundaram had by this time married Girja, daughter of the General Manager of Martin Burns Limited, a British company which was engaged in mining magnetite ore in India. Although Sundaram's salary was a small sum compared to the affluence of Girja's family, education was what had drawn the young bride. Intellect wins over material prosperity.

The years at Karaikudi were heady, but all times change, both the good and the bad. Nothing remains permanent or is forever. A carefully preserved newspaper clipping in the files of Sundaram reads: Alagappa dies at the young age of 48. The year: 1957. For Sundaram it was a deep personal loss. It was as if all the energy of the college had died out.

Equal if not more sorrowful was the fact that Sundaram was not permitted to submit his doctorate to the University of Madras. The university regulation stated that an affiliated college cannot grant a doctorate. It was only in 1985 that the government of Tamil Nadu accorded Alagappa the status of an independent university. In spite of the persuasion, authorities turned a deaf ear. Rules and regulations can wean out the best. The promise of geography seemed broken. Little did Sundaram know that geography held a brighter and different prospect. The silver lining in the dark cloud was a one-line telegram that Sundaram received: "Please come for a dialogue to Delhi. Train fare will be paid." Who was keen to initiate a dialogue? Why was the dialogue to be held in Delhi? Why was Sundaram selected to be part of the dialogue? To understand this one needs to grasp the promise of the country in the 1950s.

Promise of planning and planned development

The year Sundaram received the telegram, India had completed ten years of independence. But with independence came Partition and what followed was a large refugee movement across the Indo-Pakistan border. Delhi was a major recipient of immigrants. Data from the Census of India confirms that within a short span of ten years the population density in the city more than quadrupled. In trying to accommodate the influx, many squatter and unplanned settlements had mushroomed across the cityscape. When, Jawaharlal Nehru, India's first prime minister made a chance visit to the Old Delhi, the squalor and lack of civic amenities shocked and angered him. Keen to act Nehru summoned Kumari Amrit Kaur, the health minister to an urgent meeting and spelt out the need for city planning.

The British had established a Town Duties Committee in 1824, followed by the Municipal Committee of Delhi in 1883 and the Delhi Improvement Trust in 1937, yet none of these bodies were equipped to deal with the city's growing problems. Delhi needed a plan.

The Delhi Development Authority was created in 1957 and was entrusted with the responsibility of creating a master plan for Delhi, but being novice at the work entrusted to them, the authority needed experts. Aware that the Western world had grappled with problems of city slumification and sought scientific solutions, Nehru called upon the assistance of the Ford Foundation. In Ford, was Albert Mayor, the renowned planner who inaugurated the pilot development project at Mahewa in the district Etawah in 1948. While Ford Foundation was willing to pitch advice for the master plan, they too needed local partners.

In the 1950s India had no town planners. Even a curriculum that taught the principles of town planning was missing from campuses. In this vacuum what seems to have guided the selection of candidates was a committee under the chairmanship of Sir George Schuster which was appointed in May 1948 in the UK. The report states, "It should be obvious that physical, economic and human geography are all directly and intimately linked with planning and can provide the planner with a very large proportion of the basic knowledge he requires. Among the subjects suitable for study preliminary to planning, geography must hold a very high place" (Schuster 1950).This partly explains the need of geographers, but does not tell us why Sundaram was contacted; after all he was not the only geographer in India! The reason is worth an inquiry.

While serving as lecturer at the Alagappa College, Sundaram had tripped upon a research article on how to identify hinterland boundaries of New York City and Boston in the journal of *Economic Geography* (1955). This research carried a note on the use of telephone calls as a measure of the centrality of specific communities.

In the 1950s Karaikudi had only three telephones, one in the post office, the second in the police station, and the third in the Alagappa College. A telephone was obviously an inappropriate indicator for identifying the influence of a town in India. How could the catchment of Karaikudi be mapped?

Karaikudi was a shanty town, a market for villagers in the surrounding region. Sundaram observed that Karaikudi held a weekly *haat* where people from the surrounding villages sold their goods and bought their essentials. Further he noted that villagers returned home only after watching a movie show. Karaikudi had three cinema halls. Sundaram enrolled the owner of these halls to ask each person who purchased a ticket for the show, the name of the village from which he hailed. This was done repeatedly over six weeks. When plotted on a large scale map the results revealed the hinterland of Karaikudi. When the research was presented at a seminar organized under the auspices of the Ford Foundation in Bangalore, Sundaram's acumen for serious research was noticed. To date the Ford office has an uncanny knack for searching out the appropriate and the best.

The dialogue was an interview wherein the Town and Country Planning Organization proposed Sundaram to join the team working on the Master Plan of Delhi. The offer came with a one year contract and a salary twice that of a lecturer. At first Sundaram was reluctant. Not only did his parents severely resent relocation to Delhi but to give up a permanent job as lecturer was also a point to consider. As Sundaram boarded and headed homeward on the oldest and most famous train of India, the Grand Trunk Express, little was he to know that barely a month later he would be back on the Grand. A telegram requesting Dr A. Lakshmanaswamy Mudaliar, then vice chancellor of the University of Madras, to persuade Sundaram reached faster than the Express. Sundaram was back in Delhi.

A synchrony of events is often called destiny. There is little doubt that the death of Alagappa, coupled with Sundaram's disillusionment with the rules of academia were instrumental factors. But it also must be conceded that Sundaram was not only keen to explore a new contribution that geography could make but he also had the courage to resign from a secure tenure and take up employment on a one year contract. Few would dare to risk the safe and secure portals that academia offers. Few pursue dreams. Sundaram had begun to take courageous decisions. It is the likes of him that opened ways for geographers within the ambit of planning.

The relocation from south to north India gave Sundaram a type of stability. In the south Sundaram had changed residence six times – Calicut, Devakottai, Tirunelveli, Madurai, Chennai, Karaikudi – but in the north, Delhi became the one and only anchor. Though the shift to Delhi was a profound change in the direction of his career, Sundaram did not abandon geography!

Sundaram served the Town Planning Organization from 1958 to 1964. When the master plan document was completed in 1962 an emerging realization of the authorities was that town and country and urban and rural were so welded that planning for one at the neglect of the other would be counterproductive to both. Thus was created the Central Regional and Urban Planning Organization. Later this was merged with the Town Planning Organization to become the Central Town and Country Planning Organization under the Ministry of Works and Housing. In this context, the government of India in 1963 appointed a high level committee on rural–urban relations in which Sundaram served as senior research officer from 1963 to 1966. Since the area of operation of the committee was vast it provided opportunity to study the emerging issues in rural–urban linkages across different parts of the country. Hailing from a teaching post, Sundaram had sampled much of the theoretical and textbook geography yet he was still in the dark of how one moved from textbook to professional observation and understanding. Working amidst a multidisciplinary team the years with the planning bodies was a platform of apprenticeship.

By the 1960s India witnessed an upsurge in research on urban and regional development. The Indian Institute of Technology, Kharagpur, introduced a course on architecture and regional planning in 1952. The Regional Science Association was founded in 1954. Around this time in 1958 the Regional Planning Unit at the Planning Commission was created. The University of Mysore established the Institute of Development Studies in 1971. The Institute of Social and Economic Change, Bangalore was set up in 1972. The Centre for the Study of Regional Development was born in 1971 at Jawahar Lal Nehru University, New Delhi. It was a phase of urban planning. Even Chandigarh, the capital of Punjab, designed by the Swiss-born architect Le Corbusier, was planned in the 1950s, and most of the city was completed in the early 1960s. The mood of the country was positively tilted towards not just planning but planned development. It is the latter which was to become Sundaram's next arena of commitment.

In the Planning Commission

To lay down the objectives of planned development, the Planning Commission was established in 1950. The commission's target was to achieve its objectives within a five-year block. This led to the idea of Five-Year Plans. It was P.C. Mahalonobis, the famous statistician who advocated that no planning was going to succeed unless the spatial base and its modalities were significantly enlarged. This created the willingness to explore the role of geographers. Mahalonobis invited O.H.K. Spate from the University of Canberra and A.T.A. Learmonth from the University of Liverpool. Under the Technical Cooperation Program of the Colombo plan, Learmonth put together a team of economists, statisticians and geographers to undertake regional surveys for purposes of planning; Prakasa Rao was among them. While he was no longer Sundaram's doctoral supervisor, he continued to be a mentor. Such were the bonds that when Prakasa Rao joined the Regional Planning Unit, Indian Statistical Institute and later became chairperson of the Department of Geography, University of Delhi, Sundaram went alongside. To share the teaching load of regional planning with Prakasa Rao, Sundaram was guest faculty in the Department of Geography, University of Delhi. Similarly in 1967–1968, when the Planning Commission granted a research project to Prakasa Rao on the factors of growth in Muzaffarnagar district Sundaram was an active contributor to its fieldwork and data analysis.

These were the ways in which Sundaram remained constantly in contact with the latest research in the parent discipline. This interaction allowed Sundaram access to update his skills, tools and concepts, which became an asset to planning. The exchange of experiences from the planning organization into the geographical forums was also enriching. Sundaram was well aware of the value of this two-way relationship. He believed interdependence is a far greater virtue than independence and thus practised it.

At the beginning of the Fourth Plan when regional planning made its debut, the Planning Commission advertised a post for a geographer. Sundaram was the obvious choice. In 1973, he was appointed joint director, with special responsibilities for multilevel planning. From there on he held various positions as director, deputy adviser and then joint adviser of the multilevel planning division of the Planning Commission.

Joining a formidable national level institute is no doubt a professional asset but to carve a niche is equally challenging. Fortunately, a timely note on multilevel planning circulated by Professor Sukhomoy Chakravarty, Member, Planning Commission stated that

> the concept of multilevel planning is based on the desire to reconcile decentralisation with efficiency…problems of regional imbalances could be corrected through more decentralisation of financial responsibilities in plan making and implementation…In order to ensure that the required materials reached the local level in right time, it was necessary to consolidate the demands of the States and the districts and correlate them and ensure that these were made available
>
> (Planning Commission, 1972)

This note became the nucleus around which were spun many ideas of spatial planning.

Sundaram served the Commission from 1973 to 1986 and these 14 years rolled across the Fifth, Sixth and two years of the Seventh plan. The nature of the plans was more than numbers. The development strategy followed in the Fifth Plan had reflections of the growth centre and central place strategies. The Sixth Plan retained the concept of district planning. Sundaram was a contributor to all and more. When the Planning Commission thought that a relook at district planning was overdue and constituted a group on district planning with Professor C.H. Hanumantha

Rao as chairman, Sundaram was an active member. These works were the springboard from which Sundaram jumped headlong to join the league of international organizations.

With the international organizations

Being the premier think tank of the nation, the Planning Commission was a space where many international partners of development met frequently. The visibility of Sundaram's works opened the door to working with international organizations. The premier was the United Nations and its varied organs.

When Sundaram was working with the Central Regional and Urban Planning Organization on the Dandakaranya, a region of forest-based tribal economy in central India, he received an opportunity to visit the United Nations Research Institute for Social Development in Geneva. Sundaram was one of the first resource persons from India to be invited by the United Nations Centre for Regional Development in Nagoya, Japan. Not only did this centre appoint Sundaram as consultant but also engaged him from 1973 to 1974 in its project on the role of medium-sized towns in promoting rural development. This was followed in 1982 with a comparative study on regional development alternatives in rural societies. The United Nations Centre for Regional Development forwarded Sundaram's name to study the problems of regional development in the Philippines. When the UN set up a local office in Delhi to provide services of regional planning to various neighbouring countries, Sundaram was put in charge. He was also invited by a string of organizations which, though autonomous, drew their basic guidelines and funds from the United Nations

In 1993, Sundaram was on a month long consultancy with the Asian Institute of Technology, Bangkok for rural–regional development networks. When the United Nations Capital Development Fund and Asian Institute of Technology, Bangkok, organized a workshop on decentralization, local governance and rural development in 1999, Sundaram was the policy expert for sharing cross-country experiences. The International Centre for Integrated Mountain Development was established in Nepal in 1984, and in 1994 it used Sundaram's services for preparing manuals to train district level workers. For nearly six months in 1992 and also in 1995 Sundaram was a consultant with the Centre on Integrated Rural Development for Asia and the Pacific, an institution for promoting integrated rural development in the region.

Evidence of the fact that Sundaram left an indelible stamp is that these organizations invited him again and again. Sundaram went to United Nations Centre for Regional Development four times; to the International Centre for Integrated Mountain Development three times, and twice each to the United Nations Economic and Social Commission for Asia and the Pacific, the Asian Institute of Technology, and the United Nations

Educational, Scientific and Cultural Organization. Whereas a first time can be accounted to chance or social networking a repeat is usually based on performance.

Sundaram's passport carries a one-liner in French *Nations Unies Laissez-passer*, project manager Food and Agriculture Organization of the United Nations. In 1983 Sundaram was invited as a resource person on multilevel planning in Bangkok. Thereafter Sundaram recommended that the Food and Agriculture Organization should make a critical assessment of the experience gained in multilevel planning in the countries of the Asian and Pacific regions. When the responsibility to prepare a monograph was entrusted to Sundaram he toured six countries in Asia to collect the relevant documents. The temperament and zeal of Sundaram so matched the style of Food and Agriculture Organization that Sundaram decided to join them on a more continuous basis.

When Sundaram submitted his resignation, Dr Manmohan Singh was Deputy Chairman of the Planning Commission at that time. The latter was reluctant to part with a colleague so earnest and hard working. But Sundaram's decision prevailed. It is a different matter that even Singh resigned from the Planning Commission and joined the Food and Agriculture Organization a year after, but of course at a much senior post.

During his stay at *Viale delle Terme di Caracalla*, the Food and Agriculture Organization headquarters, Rome, Sundaram initiated several studies relating to multilevel planning with special focus on the development of small farmers.

As a geographer–planner, Sundaram had thus climbed many rungs on the ladder of success. From a small office of town planners located in Old Delhi, where the Delhi Master Plan was drafted to the Central Regional and Urban Planning Organization, which had a pan-India agenda, to the formidable Planning Commission and then on to the United Nations, Sundaram covered a large swathe. Geography delivered the promise. Sundaram drew strength from a training that was worldwide.

Learning planning worldwide

An exposure and experience drawn from the world outside of India gave Sundaram ample opportunity to grasp the nitty-gritty of planning. Sundaram was sent for a month of training on physical and town planning organized by the German Foundation for Developing Countries in Berlin. The year Sundaram visited Germany, Walter Christaller was busy working with Baskin on the translation of his book Central Places in Southern Germany. So enthused was Sundaram that he took special time out to meet the scholar at his residence at *Jugenheim an der Bergstrasse*, Germany. The meeting was memorable and the generous Christaller gifted Sundaram a manuscript of the forthcoming English translation. It indeed was a prized possession which Sundaram kept for 40 years. Today it occupies a place of pride on my bookshelf…after all it is a copy autographed by Christaller!

Beside Germany, the Union of Soviet Socialist Republics (USSR) was another country that vigorously promoted the concept of planned regional development. In 1978 at the time when I.K. Gujral was the foreign minister of India, the Indo-Soviet Commission held a symposium on planning in the field of social science. Sundaram was invited to take part in the deliberations. While residing at Tbilisi a city of Eastern Georgia, Sundaram recalls the many seminal presentations on planned development. But more than Germany and the USSR, the country and people which were to have a profound impact on Sundaram's intellectual growth were the British.

The first Indo-British seminar was held at the Department of Geography, University of Delhi in 1972 to reflect upon various issues of rural–urban interaction. So impressive were Sundaram's presentations that when an exchange seminar was held, at the University of Cambridge, in 1975, Sundaram was an invitee. Although in Cambridge for a brief fortnight, this angular city full of geometrically correct buildings, manicured squares of lawns and the River Cam won the heart of Sundaram. The tone set by academia was a most fascinating experience and Sundaram hoped that he would be a part of it someday. Dreams manifest when honed by work.

An invitation for prestigious post-doctoral research as senior Nuffield fellow in 1978–1979 provided the opportunity to be at the Centre of South Asian Studies, University of Cambridge. Most rewarding were the days spent in the immensely rich library, the hours attending the Kingsley Martin Lectures and the evenings spent at the high table dinner a tradition of Cambridge, designed to encourage interaction between scholars.

Cambridge opened the way for a series of invitations across Europe and the United States of America. As Sundaram gave lectures at the universities of Swansea, Birmingham, Utrecht, Bordeaux and, across the Atlantic, in Boston and Michigan, he gathered a precious bouquet of insight on planning and development. It is interesting to note that the majority of Sundaram's interaction in the Western realm was with geographers, a fact which allowed him to absorb many spatial dimensions about planning.

The international experience was not just about lessons of development but provided Sundaram a global taste. Sundaram's favourite cities remain Nagoya, Rome and Cambridge. The mementoes that decorate his living room include the Colosseum in Italy, pyramids of Egypt and the Hiroshima Peace Memorial, Japan. Michelangelo's sculpture in Rome, the Ikebana of Japan and the English landscape are treats which Sundaram can never forget. While the West was a region of appreciation and erudition the East became a realm of relearning and unlearning. While in the former, Sundaram studied the dimensions of planning, it was in the latter that he tried to apply and deliver the fruits of development. This geographical shift allowed him to grasp the shortfall between theoretical and applied knowledge, and thus opened doors to the possibility of a more lasting contribution.

Sundaram's main tracks took him several times to Western countries, but if the 2,684 days he spent abroad are noted from his passport, the distribution stands as follows: Europe 530, USA 82, Africa 240, and Asia (excluding

India) 1,832 days. Though Sundaram travelled far and wide, nearly 75 per cent of his foreign visits were to the under and lesser developed parts of the world, including countries like Ghana, Uganda, Bangladesh, Malaysia, Nepal, Philippines and Thailand. Even in Japan it was the third world countries that engaged his time. The type of countries Sundaram served confirms that he truly was committed to the cause of wanting to deliver planned development. Take Ghana. Not only is 60 per cent of the population poor but other indicators are just as worrisome. Life expectancy is 54 years, infant mortality is 77 deaths per 1,000 birth, 30 per cent of the households are the poorest and school enrolment for age 6 to 10 years only 57 per cent (World Bank 1990).

Be it in the capacity of a consultant, chief adviser or on deputation to conduct training programmes, such was Sundaram's contribution that most countries sent in repeated requests for him. The neighbouring countries were the most demanding. Bangladesh invited him twice, once in 1992 and again in 1995. Serving on different projects he was in Nepal three times in 1988, 1990 and 1993. He visited the Philippines twice, once in 1973–1974 and later in 1987.

This worldwide sojourn became an occasion to work with both autocrats and democrats. Brezhnev was the general secretary of the Communist Party when Sundaram was in Tbilisi, Georgia. One evening Sundaram decided to take a walk along Kura, the river on whose banks the city of Tbilisi is built. He had barely walked a few paces when guards urged him back to his hostel. Sundaram recalls a similar incident of being constantly watched while he ventured out of his hotel in Naga, a city in the Bicol region of the Philippines. Sundaram's personal meeting with Ferdinand Marcos, the president of the Philippines, amply confirmed Marcos' dictatorial style. Likewise, when Sundaram was invited by George Cosmas Adyebo, who was the ruler of Uganda following Idi Amin, the country was still carrying the imprints of fear. Sundaram was confined to Adyebo's palace and could venture out only after seeking formal permission.

In contrast within these restrictions which also meant that many reports penned were either modified or never released to the public, Sundaram's experience in his own home country India had allowed him to work within a democratic environment. The Prime Minister of India is also the Chairman of the Planning Commission. Sundaram thus served four prime ministers. Witnessing the sea of generation change from Nehru to first Indira, then Rajiv Gandhi, Sundaram also had the fortune of working directly with Prime Minister, Dr Manmohan Singh.

As a geographer–planner Sundaram had thus acquired a worldwide traverse. Starting out as a lecturer in a college, he shifted gear to become an officer in a government organization, and then rose to the rank of a division chief in a national institution, followed by tenure with renowned international bodies. He thus spent seven years teaching, 15 years in planning, 13 years in development planning, and 12 years as a consultant with international agencies. From teaching into research, followed by the bureaucratic turf of policy corridors Sundaram's graph of work displays a diversity and ascent that is uncommon to geographers in India. But more than posts, politicians and countries, what matters in the final reckoning is the work. It would be worthwhile at this juncture to attempt an audit of his work.

An audit

It is difficult to put on record the works of any eminent professional. After all how does one account for notes written, protocols kept, meetings attended, conferences arranged, drafts redrafted, letters mailed? How does one document the advice proffered, the knowledge shared and the books read? How would one even measure the hours spent reflecting on theories, models and modules? The heaps of typed papers, thanks to the support of endless efficient secretaries, are proof of the pace at which Sundaram reeled out ideas and notes. But there remains no record of the meeting notes and fieldtrips made nor are there copies of his mammoth correspondence. However if the question, "How does Sundaram work?" were asked, a simple and honest answer would be, "All the time."

If a person works all the time his stocks are high and hard to put into a ledger. Bacon once wrote to the effect that reading maketh a full man, conference a ready man and writing an exact man, Bacon's statement in his essay, "On Studies."

Bacon once wrote to the effect that reading and study made a full man; he added however that speaking makes a ready man and writing an exact man. Sundaram wrote profusely. An accessible and authentic yardstick is publications. In spite of the fact that Sundaram was not under pressure to publish or perish, prevalent across universities of meaningful standards, he has a large number of published and unpublished works.

Most in academia would keep a meticulous record of each and every piece of their publication; in contrast Sundaram's curriculum vitae is shockingly incomplete. The route to piece together his publications turned into a treasure hunt. As one moved from clue to clue, books and articles were relatively easier to retrieve; the hard part was the reports. This is because the ownership of reports is claimed by sponsoring institutions, and even if an individual has made a substantial contribution, the credit is rarely made visible. Even though an incomplete list, his large stock carries reports, monographs, training modules, reviews, research books and articles.

What were Sundaram's most prolific years? Before an assessment can be drawn of the kitty of Sundaram's works there is a need to create an equivalence scale. After all, ascribing equal weight to the publication of a book *vis-à-vis* an article in a journal would be like considering apples and oranges as a similar fruit. The method I have devised is the conversion of all forms of publications be it a book, report or article to what I call a research unit, the idea being that the assortment is converted into a standard unit. Sundaram's varied forms of publication are converted into a single research article. Thus an article in a journal is counted as one unit, similarly a chapter in a book is also treated as one. To calculate the units in a self-authored book, the latter is divided into the number of chapters it contains – a book with six chapter means six research units. In an edited book only the chapters written by Sundaram were considered and each given a value of one unit. Reports were more ticklish, but considering that institutes of eminence like the Planning Commission or Food and Agriculture Organisation are the publishers, they were ascribed a unit value of three. Using such a conversion, Sundaram research units became a sizeable pack of 281.

When viewed on the graph, Sundaram's output of work is impressive (Figure 5.1). The trend is upward. The line is consistent and has no breaks. There is a publication each and every year.

Beginning from the young age of 31 when he launched his first piece of writing well up to the age of 80, Sundaram has published consistently and continuously. Interestingly his period of maximum research units was when he was 68 years of age. This year alone he published 24 research units. Sundaram added a dozen research units annually to his graph during the 1940s and 1950s. The most prolific decades were the 1980s and 1990s in which Sundaram wrote 62 per cent of his works.

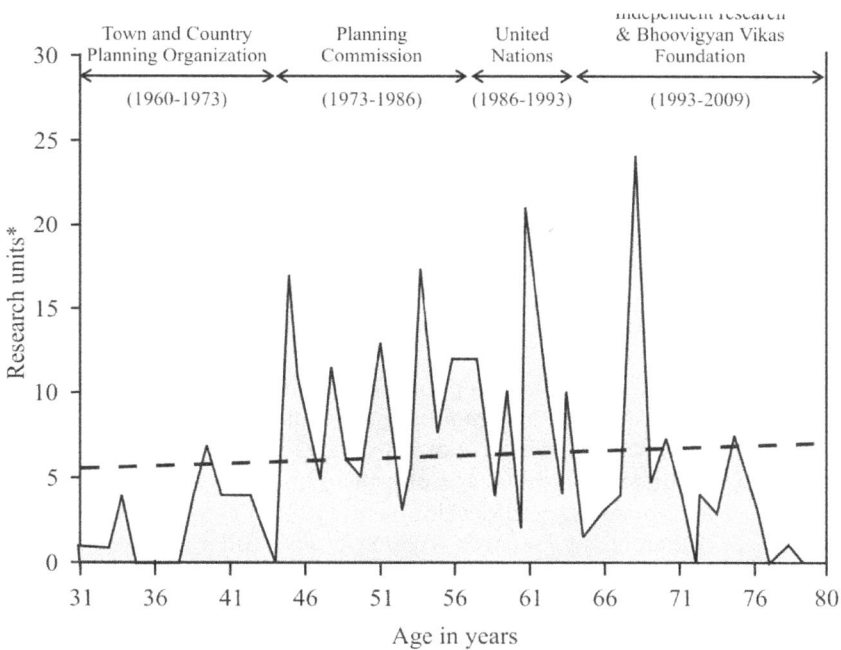

Figure 5.1 An Audit.
Source: Author.

Seen against the backdrop of the organization with whom he was working, the years at the Planning Commission were most productive and contributed a good 50 per cent of his research units.

There is little doubt that outside the precincts of the university Sundaram was the most productive geographer in India. Even when compared to those engaged in full time academics, Sundaram's publications add up to more than 100 geographers in India! In the intensity with which he lived and worked Sundaram stands a class apart. He wrote for students, researchers, trainers and the bureaucracy. Apart from English his works are translated into Nepalese, Japanese and Russian. But in work and thinking Sundaram is a loner. Of the total publications, 237 (84%) have been written alone. Almost all his serious research publications are single authored works. However, he is generous and accommodative. Nearly 32 of his works are co-authored and 15 are with more than one author. The latter belong to as diverse disciplines as architects, agricultural scientists, economists, sociologists and environmentalists. That he never abandoned his parent discipline is confirmed by the fact that geographical journals like the *Bombay Geographical*, *Annals of the National Association of* Geographers, *India*, and the *Analytical Geographer* carry his articles. Sundaram has co-authored/edited ten books with colleagues in various geography departments of India. Moreover, when a sample from four of Sundaram's edited books was culled, 50 per cent of the 110 authors in these books turned out to be geographers.[1] But more important than sheer numbers is the intellectual interest he nurtured.

Intellectual interest

Among the infinite variety of spatial problems that make life worth living as a geographer Sundaram chose regional planning and development. In his resume he calls himself a regional development planner. The thousand odd books in his personal library are dominated by titles on development and planning be they urban, agriculture, rural or regional. But the true evidence of his quest to contribute to planned development rests in the titles of his very own works.

When the gamut of Sundaram's published and unpublished works including his important speeches is compiled they form a sum total of 352 titles. A thematic classification of these words provides an interesting insight into what engaged Sundaram's mind. The classification falls into three neat compartments: area, planning and development. Each of these acquires a near equal share. References to an area whether block, district, region, rural or urban, constitute 29 per cent of the words in the title. This is followed by the word development which represents another 28 per cent. Be it decentralization, multilevel, *panchayat*, participatory or spatial, the word plan or planning occupies a sizeable 20 per cent of the words in the title. The remaining words were either additional specificities like agriculture, land use, environment or carried philosophical overtones like a concept, approach, problem or trend. Therefore reflected in the very titles of Sundaram's research work is a consistent commitment to an area and its planned development. But Sundaram's concern is not static. As he tries to capture so elusive a concept as planning or development so too do his publications change in content and complexion. This becomes apparent in a snapshot review of three of his most outstanding works.

The first is *Urban and Regional Planning in India* (1977). Of its 21 chapters, 12 are purely on urban and settlement systems, three are on rural–urban relationships, two each deal with tribal and regional issues while there is only one chapter exclusively on rural areas. The crux of this book, whether rural–urban interaction or the community of interest area or land use, revolves around urban. The focus on urban is the natural outcome of Sundaram's 15 years at the Town and Country Planning Organization. Exploratory in content, backed by in-depth case studies and rich in data base in this 437 page book Sundaram displays a deep influence of the models of Western thinkers such as Christaller, Hoytt or, for that matter, Harris and Ullman. While Sundaram is entirely drenched in urban in the 1960s and the 1970s he acquires the ability to swim out of it and into the regional only after he has spent a good number of years grappling with the issues of disparity and uneven spatial development.

His second book, *Geography of Underdevelopment: The Spatial Dynamics of Underdevelopment* (1983), is a work that shifts from an Indian to a global perspective. Undertaking an in-depth review of literature on efforts of development across the world, this book of ten chapters concentrates on issues of spatial underdevelopment across various countries. Sundaram's mindscape, as mirrored in the book, is one that is hesitant and filled with pensive thoughts. The heightened optimism in planning that is reflected in the chapters of *Urban and Regional Planning* is clearly missing. Sundaram makes great effort to piece together the feeble examples where planning for development has been able to make a dent. But Sundaram is neither a pessimist nor an optimist; taking hold of the lofty promise of planning he indulges to experiment and eke out a solution this time not on the urban, nor on the vast regional, but on the rural area.

Sundaram's third milestone is *Decentralized Multilevel Planning: Principle and Practice, Asian and African Experience* in which he shares the experience of decentralization across Africa and Asia. Published in 1997, it is a work that came to light after Sundaram spent 14 years as a consultant with international organizations. The book belongs to Sundaram's reflective phase. The 20 chapters of this 514-page book are about rural areas and attempt to reach out to the village and the farmer. The work in fact does not have a single chapter on urban areas. The book is universal; it is not focused on India but addresses all countries that are grappling with issues of how to bring about development.

Backed by in-depth dos and don'ts the prime message is that rural development can be achieved if appropriate modes of decentralization are adopted and allowed to function. Adopting a Gandhian philosophy, self-reliance is seen as the basis of decentralization. According to Gandhi, a viable political system for India has to be centred on village republics, organized like oceanic circles. The metaphor was meant to convey the principle of local power in combination with commitment to the larger society. The cover jacket of this book of Sundaram carries a graphic of the oceanic circles advocated by Gandhi.

Punctuated with these three landmarks, Sundaram's publications evolve from the urban to the global to the rural, from case studies of cities to countries, from the Western to the indigenous Gandhian model.

As Sundaram's search for answers and solutions continues, he claims he has been on the Trodden Path. Beefed up by a number of autobiographical pieces that provide narrative as to the way Sundaram wants to be understood, the contents of the book *Trodden Path* draws chapters from earlier publications. But what is interesting is its cover page: a man with a pensive gait heading towards a vast and distanced ruralscape. Is development yet so distant from the vast ruralscape? Is it the trodden path of Sundaram? Is it the path that won him awards and rewards?

Awards and rewards

Sundaram's work earned him repute. Shining like medals, the walls of Sundaram's office, at his residence, are filled with placards, shields, plates and mementoes. They are a sum of over two dozen and have been given by different types of organizations, across different parts of the world. It would be difficult to list all of them here, but some deserve a mention. Sundaram received a placard of appreciation from the German Foundation for Developing Countries (now called the German Foundation for International Development) and the government of Naga, Philippines in 1970 and another book in 1973. Half a dozen felicitations were received from centres like the Asian Institute of Technology, Bangkok; Banaras Hindu University, Varanasi; the National Institute of Rural Development; and the University of Delhi. The National Association of Geographers, India elected him as president and congratulated him for the active and inspiring tenure of his presidency marked by notable innovations and contributions to the cause of geography. Sundaram's name is listed in India's "Who's Who." He is also included in the *Learned India : Men and Women of Achievements and Distinctions* and is among those with a "Distinguished Leadership Award and World Decoration of Excellence." He has been invited to deliver eminent lectures such as the first Prakasa Rao memorial lecture (1997); the R.N. Dubey memorial foundation lecture (1999); the S.P. Chatterjee memorial lecture; and the N.P. Ayyar memorial lecture (2001). But the jewel in the crown is the Jayendra Saraswathi Lifetime Excellence and Achievement Award bestowed on Sundaram by the Centenarian Trust, Chennai (India), for, "…outstanding geographer–planner of India … to mark singular dedication to the cause and care of the earth." Among the list of recipients of this lifetime award are no less than Dr Manmohan Singh and Dr A.P.J. Abdul Kalam.

While these are awards, the true reward of a professional is recognition of his work. Sundaram did not lack on this front. From his formative years Sundaram proved that scholarship is a virtue that defies degrees. Sundaram moved directly from an undergraduate degree to a D.Litt. In fact his portfolio of certificates carries only two university degrees.

Dated 19 August 1955, on his graduation degree is inscribed:

> The Senate of the University of Madras hereby makes known that K.V. Sundaram who was admitted to the degree of Bachelor of Science (Honours) at the Convocation held in the year 1953....is now without further examination and by virtue of Regulation 2 of Chapter XLV of the Laws of the University is admitted by us to the Degree of Master of Arts.

Jumping a grade may seem an official ease of academia in the 1940s, but the next jump is truly commendable if one recalls that Sundaram was not given permission by the University of Madras to submit his near completed doctorate. Yet his name carries the prefix Dr. The second degree states:

> We the Chancellor, Vice Chancellor and the Senate of the University of Mysore do hereby certify that Sundaram, K.V. has been admitted to the Degree of Doctor of Letters (D.Litt.)... having been duly certified to have passed the examination prescribed therefore in the year 1982.

In the UK, Australia, India and certain other countries, the degree of Doctor of Letters is valued higher than that of Doctor of Philosophy, and is bestowed on the basis of a long record of research and publication. It was Sundaram's book *Urban and Regional Planning in India* (1977) which earned him the D.Litt. degree from the University of Mysore (India). The book was widely reviewed and leading national dailies like *Hindustan Times*, *National Herald*, *Financial Express* and *Deccan Herald* carried a piece on it.

The basis of the second book, *Geography of Underdevelopment*, is also praiseworthy. B.H Farmer, the director of the Centre of South Asian Studies, University of Cambridge had a special interest in India and Sri Lanka (then Ceylon) and therefore developed close contact with geographers in these countries. In 1972, it was at the Indo-British seminar held in Delhi that Sundaram impressed Farmer. Like all committed academicians who are keen to build their department, Farmer had kept a watch out for Sundaram. When the latter was in service at the Planning Commission, he received the Senior Nuffield Fellowship for research at the Centre of South Asian Studies in the University of Cambridge. The foreword to the book carries a note by B.H. Farmer, the director which states, "...I hope that *Geography of Underdevelopment: The Spatial Dynamics of Underdevelopment* will have the very considerable intellectual and practical impact that it deserves." It did. The first edition was sold out very soon. Nothing gives an author greater pleasure than to find his book is out of print.

With these as the star attractions, there are many others that merit consideration. Sundaram presented a scheme of planning regions of India based on the principles of food self-sufficiency and ecological balance. This scheme of regions was incorporated in chapter nine of the third Five-Year Plan document. His report on Kundrakudi was prescribed as essential reading in the training programmes organized by the Lal Bahadur Shastri Academy of Public Administration, Mussoorie. It was also included as reading material in several training programmes on decentralized agricultural planning in Asian and African countries.

Today it seems almost impossible to visualize a geography syllabus without a course on urban and regional planning. A curriculum, however, begs for textbooks. In the 1970s there were few peers who could compete seriously with Sundaram where books on urban and regional planning are concerned. Even stalwarts like Prakasa Rao and R.P. Mishra were co-authors with Sundaram. Therefore it is safe to consider that well over a generation of geographers in India were trained on publications of multilevel and regional planning which were either authored by Sundaram or at least quote Sundaram profusely. In many syllabi across geography departments in India Sundaram's books are reference readings. Take Banaras Hindu University, the syllabus of the geography department lists Sundaram's works across three courses: rural and urban planning, resource planning and regional planning. It is no coincidence that when the Food and Agriculture Organisation forwarded a list of candidates shortlisted for consultants in rural development of Nepal to the late King Birendra, the latter promptly ticked Sundaram's name. And why not? Crown Prince Dipendra, the king's son had taken an undergraduate course of geography at the Banaras Hindu University where he had read Sundaram's books. Name builds fame.

The fact remains that the contribution of Sundaram in, "... presenting an integrated picture of multilevel planning and rural development cannot escape the attention of serious students of multilevel planning" (Mukherjee 1990a). It was those like him that played a critical role in converting the area development unit within the Planning Commission into that of multilevel planning.

Sundaram wrote before the word processor appeared and amplified the scholars' capacity for better or worse expressions. He achieved what he did the old-fashioned way, by dint of hard work. Following strictly the famous grammar text by Wren and Martin, much of what Sundaram has written is in clear and simple language. Though it lacks on the anecdotal side it also avoids the use of jargon. Since Sundaram had long served in the corridors of bureaucracy his works tend to be prescriptive but within the air of caution there is a fluency and clarity. Being a frequent visitor at all odd hours to the Sundaram residence here is the image I can draw...the aroma of South Indian coffee, Sundaram on the sofa with a clipboard and papers, filling in the pages in his surprisingly miniscule script.

It becomes plainly obvious that Sundaram had inherited a high set of standards from his father but it is also clear that he has worked to maintain those standards. If a list of prominent geographers in India to date is drawn, or a list of planner-geographers or contributors to spatial policy, they all will surely contain Sundaram. He was a master geographer–planner. It would sure be worth asking what does a master do? How does a master be!

Master ...

Masters...Learn from disappointment and success

Cumbum is a valley in the Theni district of Tamil Nadu near the Kerala border. It is in this wedge-shaped valley that Sundaram spent nearly a decade gathering data on its agricultural characteristics for his doctorate. But when the rules of the University of Madras framed in 1860 did not recognize a doctorate from an affiliated college Sundaram writes, "I tore to pieces all my research papers ... so that no shred of evidence of any kind was left about my research in the Cumbum Valley." That this early professional disappointment continued to stay with Sundaram is apparent from the fact that he narrates this trauma even in his penultimate book *Trodden Path*. But neither does shredding a manuscript erase memory nor gaining a DLitt make up for the years of sadness. This was not the last of his struggles. If academia has its share of troubles, other professions are no rosier.

There are two broad groups of planners in the country. One consists of socio-economic planners and the other is a more organized or institutionalized group of physical planners. The Delhi Master Plan was an interdisciplinary effort where 28 physical planners collaborated with 17 economists, four geographers and three sociologists. Hesitant to state the negative, Sundaram writes, "undue aggression, rivalry or unconscio,us clash of personalities can destroy group effectiveness... my argument[s] for [a] greater role of social scientists in city planning were not relished by some senior town planners" and, "it must be remembered that the mere bringing together of various specialized disciplines under one roof does not lead to interdisciplinary teamwork .... You cannot buy a research team. With luck, it grows making its own common language and thriving on personal interplay which has nothing to do with research." Though luck was obviously not with Sundaram, yet later years proved that many practical problems which arose in the implementation of plans could have been avoided had community involvement been included.

Sundaram's passion to make the deliverance of development a possibility compelled him to examine cases of disappointments and success. The North-East Polder in Holland, Jezreel Valley in Israel and the Thal Canal Project in Northern Punjab in West Pakistan are examples he drew upon. Evident in his writings are the lessons also drawn from the Saemaul Undong New Community Movement in South Korea, Mao's clarion call in Tachai, China, Muhammad Yasin's YUNUS cooperative movement in Bangladesh, the development experience of Mezzogiorno subregion in Southern Italy and the Muda Irrigation Project in Malaysia.

In India, Sundaram's reviews span from the Sukhomajri, Chandigarh to the integrated rural development programme implemented at Kazhakkoottam Block in the Trivandrum district of Kerala. The role of the small farmer's development association in the Uthiramerur Block in Chingleput district, Tamil Naidu and that of the role of a leftist government in initiating decentralized planning at Midnapore, West Bengal is also scanned. After careful reviews of all such cases Sundaram concludes where and why development stumbled and fumbled. But armchair work was obviously not the style of Sundaram. To grasp the success and failures of development masters travel far and wide.

When Sundaram was on an assignment to Bangladesh in 1992 he made it a point to drive down the Chittagong-Dhakka highway to meet Akhter Hammed Khan at Comilla. A city in south-eastern Bangladesh, Comilla was selected as a pilot project site by the Pakistan Academy for Rural Development and Dr Akhter Hammed Khan was the founder director of the academy in 1959. Sundaram realized that the success of Comilla lay on the personal vision of Akhter which focused on cooperatives, irrigation and rural works within the selected blocks for planning. Sundaram's meeting with Akhter was enriching, yet more insightful was the study he undertook back home in Kundrakudi, a typical South Indian village in a drought prone part of Pasumpon Muthuramalingam district in Tamil Naidu. The opportunity came the way of Sundaram when the late prime minister, Indira Gandhi, read a report of the success of rural development in Kundrakudi in the newspaper *Hindu* in 1984. After going through this press report she directed the Planning Commission to study this success, adding, "this is what I should like for all other villages in the country." The Planning Commission appointed Sundaram as the leader of the team to document the Kundrakudi experience. When he returned and went to see Gandhi she asked, "Is it replicable?" Sundaram replied, "it is not." He explained that the moving spirit behind Kundrakudi was a spiritual leader called Adigalar who had launched a village planning forum to achieve self-sufficiency and eradicate unemployment in the village. Whereas Adigalar could not be cloned, Sundaram shares that it is the mix of three main groups of factors: technocratic, academic and democratic which makes possible rural development. Coining the word TADMIX, when the Planning Commission published the study which Sundaram wrote, it gained wide popularity within the offices of government.

With a wide spatial compass and a deep concern Sundaram's first task was always to understand the problem in its entirety. He adapted his mode of inquiry according to the objectives at hand. Like a true master he more often than not set out to inquire about many a problem through first-hand inquiry. Evident in many of his works is the role of fieldwork.

...Works in the field

Accepting that geography is a realistic science and surveys are the backbone of information that influence the proposals to be included in a development plan Sundaram indulged in fieldwork across many parts of India.

Before drafting the chapter on rural Delhi in the preparation of the First Master Plan, Sundaram undertook the survey of no less than 300 villages. The study of Kotla Mubarakpur covered a sample of 139 families in three villages, while in Meerut around 100 shops were covered to provide information about the role of this town as a regional centre. When attempting to assess the service standards of 100 municipalities across India, Sundaram embarked on an ambitious project to gather data through mailed questionnaires to the concerned departments. In Kundrakudi, Sundaram took the opportunity to sample the respondents who had been beneficiaries of development. Against official advice, Sundaram lead a team to meet up with the tribes of Bastar. The trek covered Abujh Marh, the unknown highlands. But Sundaram was undeterred. He also ventured into Morena, a dacoit and tribal infested area, and also the Dandakaranya, a place which means the jungle (*aranya)* of punishment (*dandakas*).

Keen to gain first-hand knowledge, except for the North-East, Sundaram has visited all the remaining 35 states of India. Even on his visits abroad, be it Bicol, Uganda or Ghana, Sundaram maintained a meticulous diary taking note of the interregional inequalities as well as the struggle of the governments to cope with this problem.

Whereas primary data gave Sundaram a much-needed feel for the people and the land, Sundaram did not shy away from using secondary data where the area was either too large or the problem at hand demanded a tier of quantification. The Census of India, Five-Year Plans, the National Sample Survey, Indian Statistical Institute or reports of national and international organizations among others, are frequent sources of data in his publications.

During 1965–1970 when seeking a solution to the problem of regional disparities, Sundaram identified the less developed districts in India, using a set of 14 indicators, all picked from the basket of secondary data. In an attempt to cut through the persistent regional imbalances throughout India, the Gadgil formula was devised for the allocation of central assistance to the states giving weightage to the backward states. In this context, Sundaram examined the levels of development in various states employing a set of 13 variables. The same analysis was repeated to examine the intrastate disparities in Andhra Pradesh using district level data. To determine if disparities had widened or narrowed two sets of data, one from 1957 and the other from 1965 were used.

Elaborate statistical analysis was not Sundaram's forte, yet their use in his research confirms that he did not hesitate to ask, be taught and apply. Correlation was used to find the relationship between population and net sown area or basic and non-basic employment. For resource-based regionalization in India he used the multivariate analysis. To identify the regions of development, the principal component analysis was employed.

While data and statistical techniques are important, Sundaram believed that maps are an end product of analysis. In his parade ground voice he often claimed that a map is a scale reality. There is hardly any publication of Sundaram which does not carry at least one map even if it only that of the location of the area. While his fondness for maps remained, it seems that the lack of cartographic facilities in many of the institutions he served could have been a deterrent which explains their paucity in some of his later publications.

Whereas data, techniques and maps were no doubt helpful tools, Sundaram's mindset was most fascinated with models. Perhaps he was in search for the ideal model which could lead the way to development. Strewn across Sundaram's many publications are theories and models. Here it would be pertinent to remember that from the 1970s until the mid-1980s geographers in India were engaged with testing the validity and application of spatial models of development proposed by the West. Sundaram was no exception. A person after all is a product of his time. Masters test theory.

...Test theory

In studying the morphology and characteristics of the towns of Madurai, Moradabad, Tiruchirappalli and Coimbatore, Sundaram used the sectoral, concentric, and multiple nuclei models of the internal structure of the city proposed by Burgess (1925), Hoyt (1939), and Harris (1945). In the late 1960s and early 1970s, academic discussions centred on the role of growth poles and growth centres in regional development and planning. Whenever Sundaram refers to the bottlenecks or potential for development, the growth pole theory of the French regional economist François Perroux (1950) and the backwash and spread effect of Gunnar Myrdal (1957), a Swedish economist find handy reference. Another concept which Sundaram used is the Kolb learning cycle, a four-stage model which is handy when conducting training programmes.

Sundaram refers and uses many different models but his all-time favourite was Christaller's central place theory. When drafting the rural development chapter for the Master Plan, Sundaram follows a hierarchical pattern of central villages, service centres and rural towns. When the Rajasthan canal colonization project was being planned Sundaram used the central place theory to identify the optimum size of settlement for the canal area. Selecting the size of economic holding, efficiency for services, and cost of provision of basic civic amenities as the criteria he concludes that, "... a village of less than 1000 population has to be rejected as uneconomical and a village of about 1200 population accepted as the optimum size." The spell of Christaller, whom he personally met way back in 1965 continues long and strong, even in 2001 Sundaram writes, "... one of the most fruitful theoretical and operationally feasible approaches to the study of urban growth is the central place theory by Christaller."

But masters do not carry blind faith. When applying the structure of city models Sundaram drew the conclusion that in the Indian context, where cities have long histories the resulting pattern is complex and is difficult to explain by any single theory alone. In the development of backward tribal areas, such as Bastar, Sundaram concluded that social development was more relevant and the spatial framework of settlement structure must be decided after taking note of the existing consumer travel patterns in the tribal society.

Masters do test before they settle for the best. Sundaram, however, was not an empiricist, nor a juggler of data. He was not a fieldworker nor was he a map-maker. Masters work for those in need.

…*Carries concern*

In a world filled with problems Sundaram could sift the significant from the important. Common to his research are questions like: How does one delimit the catchment area of a town? What should be the number, size and spacing of cities? What is the nature of rural and urban interaction? How can regional disparities be measured? What are the available resources for planning and development? How can standards and norms be prescribed for development? Can the optimum size and desirable population of a settlement be quantified? What are the mechanisms through which development can be delivered? How can people especially administration be equipped with the task of delivering development?

Spatially Sundaram has worked on a diverse scale. As if keen to explore every rung of the continuum, Sundaram has studies that range from a settlement to a hierarchy of settlements. There are works on the locality of a town and also ones on a town. Even the in-between rurban areas are not forgotten. He covered the block, district, state and the region. From the region to the state and the country Sundaram has works on each of these scales. He has also travelled to the neighbours and stretched to countries in Asia and Africa. With this sweep Sundaram has works across different ecological zones ranging from the Rajasthan desert, the highlands of the Deccan, Nepal Himalayas or Philippine islands.

Diversity however was not the issue, Sundaram was not seeking variety. With deep care, he was selecting areas, regions and people that were in critical need for planned development. His concern primarily was for those people and regions that should be the final recipients of planning. Thus when he chose Kotla Mubarakpur in Delhi, his focus was on a relict village, an area that urbanization had bypassed. With Kundewalan the spotlight was on the old walled city area of Delhi. The object was not to bring out the extent of deficiencies of housing and community facilities but to gain insight into the social and psychological facets which can contribute to building appropriate objectives for planning. Sundaram identified the catchment area of the town of Karaikudi and Delhi. The goal was not just to delimit the catchment of Delhi but to identify the community of interest area. The latter is a functional area in which the associations and social contacts of the rural people are voluntary. What explains the selection of the study of a medium size town like Meerut is the fact that it is this type of class and not the big towns that support and stimulate rural development.

When Sundaram selected the study of 100 urban local bodies, the goal was to assess which municipalities were able to function and why others were poor performers. Even when Sundaram shifts from the town or local planning to regional planning the concern was to secure for the great majority of the population a reasonable accessibility to all grades of service. This could be for the peasants residing in the colonies recently irrigated by the Indira Gandhi Canal Project, Rajasthan or for tribes like the *Abujh Marhis* in the Bastar district.

With these as the concerns, Sundaram clearly preferred and advised others to choose problems that were salient in real world terms. He was not one who raised questions based on curiosity or applied theories just to experiment. Above all he was a seeker of solutions. While more often than not, most people would find themselves at sea when trying to work through problems, Sundaram's inclination was to seek solutions. Being a pragmatic he eschewed theory for its own sake and used the central place theory, the growth pole theory or regional analysis and resource appraisal to solve real world problems. Ingrained in masters is the search for an answer to problems.

...In search of solutions

Every problem has a solution. But every solution also creates problems. One can hardly hope to solve a problem at the level at which it is created. Sundaram's evolving mind allowed him to propose, dispose and come up with possible ways to achieve the efforts of planning and development. All the solutions that Sundaram offered carried a sharp spatial lens. His planning was for areas, and in his view the region was the foundation stone for all planning. For Sundaram, geography was a way to clarify an issue and analyse a problem and propose a solution. The latter was most essential. When asked in an interview how geography had helped him in planning he replied that it offered an unmatched flexibility and cross-disciplinary landscape on which to operate. This is where he blossomed. I am tempted to call them "blossoms," simply because Sundaram has a special fascination for these flowers. Pictures of this plant are found on the side of his study table. A set of spirals which contain his notes, media clippings and photographs, carry the title *Blossoms in the Dust, Blossoms in the Wood, Blossoms are True*. It is difficult to establish where and why he picked up this association. It may be his frequent visits to Japan which celebrates the cherry blossom festival. It could well be the caption of Kusum Nair's soul-stirring book *Blossoms in the Dust* (1987), which assesses the community development programme in India. Blossoms are abundant in South India and their fragrance lingers long and dear. Whatever the reason wherever Sundaram spells possible solutions, the works carry the freshness of blossoms. Masters reflect upon ways to improve the world and leave it a bit more learned if not better.

The preface of the proceedings of the 21st International Geography Congress, 1968, on Population Geography states that

One of the most important recommendations which followed Sundaram's paper was that the geographical aspect should be taken into account when delimiting the urban tracts... it has since been decided to delimit the Census tracts in 1971, not on the concept of town groups.

The Town Group concept adopted in the Indian Census up to 1961 was thus replaced by the concept of the Standard Urban Area. Proposed by Sundaram, the Standard Urban Area provided consistent and comparable data on metropolitan and city areas and thus became an essential tool for purposes of planning. So powerful was the concept that years later it became the basis of a doctorate at the Department of Geography, Delhi School of Economics (Sharma 1998).

As a step to achieve regional development, Sundaram presented a scheme of planning regions of India. According to this scheme India was divided into 13 macro regions of which six were interstate regions. Formulating a regional policy at the meso level laid the scope for supporting a general policy of national economic development. This concept of interstate regions led to the emergence of a new geographical scale in planning. While Sundaram's scheme of regional classification was hailed as a breakthrough, equally important was his research on regional disparities of India. Plunging into a forest of parameters he threw up the conclusion that, "underdeveloped India is not a single problem region but lies scattered in diverse regions of markedly different natural resource endowments and socio-economic environments." Besides contributing solutions to problems of data collection for planning, or regionalization for policy analysis Sundaram was most keen to find routes through which development could be channelled. He thus toyed with and polished the idea of clusters, multilevel and stages.

Clusters and multi levels

Sundaram argues that growth needs to be structured at the level of clusters of villages in order to make rural development inclusive and reliant on people's participation. It was during the preparation of the plan of Delhi that Sundaram conceived of a rural development framework with service centres, on the basis of Christaller's concept. Albert Meyer called this the village cluster programme. A more concrete beginning towards this approach commenced when the Ford Foundation envisaged a village cluster, around an identified growth centre, as an intermediate area unit between the village and the block. This approach was also recommended by the Working Group on Block Level Planning, set up by the Planning Commission in 1978.

On the basis of his experience with African and Asian countries Sundaram concludes that before freezing any type of plan, a cluster of districts in different ecological environments should be selected to conduct experiments. Even for geography, his parent discipline, he urges that each department in the country should adopt a cluster of villages in their vicinity as a laboratory for continuous research. Along with clusters, Sundaram seeks a solution for development through multilevel planning. The position taken in his studies is that any attempt at defining or delineating multilevel planning must involve reference to regional planning. He states, "...much of the cultural heterogeneity is space specific...these space bound socio-economic structures call for space oriented action." In the case of Bastar, the solution Sundaram proposes is a five-tier decentralized pattern of social development nuclei, conceived hierarchically in terms of central villages, service towns, market towns, growth points and growth centres. In the Rajasthan Canal Project the proposed settlement hierarchy is the basic village, middle school village, the service town, the *mandi* town and the regional town.

Development, however, cannot be achieved overnight. Built into Sundaram's research is also the concept of stages. Thus the case of Meerut is discussed within a stage of growth hypothesis in which each stage in the development becomes a function of the previous stage. Similarly, to take the backward area development through successive steps of radical changes over time, a transition growth strategy is proposed. Democratic decentralization is also viewed as a stages approach to transformation. Sundaram affirms that India has successfully crossed the first stage of enfranchisement and has partially attempted the second or the empowerment stage with varying results. But the ethical mode of functioning which is called the third stage of democratic transformation remains to be addressed. While change is inevitable Sundaram cautions that solutions must be applied but with caution. Masters handle all with care.

... With care

Thus writes Sundaram, "the urban village has so far been left out in the urban development process for various reasons and it is bound to remain a maladjusted link in the urban matrix for some more time to come." His proposal to refurbish the village is in alignment with Patrick Geddes' conservative surgery approach, which he propounds in his book *Cities of Evolution* (1968). Seasoned with a worldwide experience even when advocating development for the tribal region of Bastar Sundaram advice is to hasten slowly and usher change with caution.

Above all wrote the great geographer Sauer in a memorable phrase, "locomotion should be slow: slower the better" (Sauer 1956). He was right. It takes time and consideration to understand places and people. But it also takes preparation and guidance. The most enduring contribution that Sundaram makes to achieve development is his training modules.

Model to modules

"Over six lakh forty thousand people, who man the district planning machinery at various levels, need training for proper preparation, monitoring and implementation of district plans, in the country." This was data provided in a report on Training for District Planning (Planning Commission 1989).

Sundaram realized that good plans do not by themselves guarantee anything unless appropriate ways are found to implement them. Sundaram was instrumental for initiating the centrally assisted training programmes for area development in more than a dozen institutions in the country. As a member of the Central Advisory Committee on Training, for the government of India, he collaborated with the Institute of Development Studies, Mysore, the Administrative Staff College of India, Hyderabad and the Institute of Economic Growth, Delhi, for running training courses. Toeing this idea he established a link between the Planning Commission and the Department of Science and Technology to set up the information systems for planning at the subnational levels. His ten-volume set of training manuals has long been a standard part of the adviser's kit for less developed countries. Sundaram shift from models to modules was an act of retooling which gave him a new and perhaps a more promising area to explore.

Since Sundaram had worked and walked to find ways to achieve development he, more than others, was aware of the many roadblocks that obstruct the way. When answers are hard to come by, masters turn philosopher and poet. Tucked in the speeches and publications of Sundaram, especially of the latter years, are frequent quotes by T.H. Lawrence, M.K. Gandhi, R. Tagore, and W. Wordsworth. Musing on the tardy pace of decentralized multilevel planning in India he uses Wordsworth to express, "… it is still a hope still longed for never seen…" In many such selections resound a philosophy of detachment but not of indifference. The likes of Sundaram can neither be pessimistic nor cynical. Masters live a life filled with gratitude.

….Live in gratitude

Following a near spiritual evolution, Sundaram dedicates his first book to his parents, the second to his wife, and the third to providence. But the one name that is repeated in each and every page of acknowledgement is Girja, his wife. Like many traditional wives in India, Girja performs innumerable chores but, unlike many others, to Sundaram, she has been a record keeper, proofreader, fact finder, organizer and a keen participant in his works. When Sundaram speaks from the podium, Girja's eyes shine with pride. Support is the need of each and every soul.

Sundaram's circle of gratitude expands beyond the family. Prakasa Rao was among those who provided the intellectual heft behind many of the works Sundaram undertook. It would be fitting to say that if studies on regional development became the focal point of research in India in the 1960s much credit goes to Prakasa Rao, but carrying the relay forward was Sundaram. Interestingly, Prakasa Rao and Sundaram carry many semblances, for example, both had an active interest in urban and regional planning. Sundaram's research on the hinterland of Karaikudi precedes his meeting Rao by a good ten years. Both were part of the University of Madras, one as student and Rao as faculty. Both left South India to move to the north. Both were associated with the Planning Commission, Rao on projects and Sundaram on a permanent tenure. Both have worked with the United Nations, Rao for a brief period and Sundaram for a long span. While Sundaram and Prakasa Rao have co-authored three books, Sundaram never forgot that Prakasa Rao was his mentor. He edited a *festschrift* in honour of Prakasa Rao in 1985. Sundaram is generous and for a two-volume book extended this gesture even to a friend and colleague, R.P. Mishra.

The Sundaram–Mishra camaraderie is well known and they also share many similarities: they are almost the same age, Mishra being only one year younger; they both hail from small towns and share a keen interest in regional development; both believe in the Gandhian philosophy of development; both served the United Nations Centre for Regional Development as well as serving as presidents of the National Association of Geographers, India; and, finally, both have co-authored three books. Incidentally, they met for the first time in 1966 on a Sunday afternoon at Prakasa Rao's residence. Today in 2010 both reside in Delhi. Their acknowledgement to each other runs deep. While Mishra edited a slim book as a gesture of regard on Sundaram's 60th birthday, Sundaram's return gift was a two-volume set of essays in honour of his friend. But beyond family, research and friendship, Sundaram's deepest gesture of gratitude remains towards the discipline of geography.

Sundaram claimed that geography had given him the ability to grasp interactions, relationships and problems within a spatial context. It was this discipline that has built his moorings of regional development. The study of geography provided the skills to create maps and conduct fieldwork, and the ability to synthesize. An integrated spatial perspective became important even to build a view of clusters, multilevels and stages. But while the subject was enriching, Sundaram was dismayed to observe its languishing status in India. As president of the largest geographical society in India, Sundaram drew up a matrix of the strengths, weaknesses, opportunities and threats of geography. He was convinced that in India the discipline desperately needed to be pulled out of its morass. Sundaram is a doer. He is a giver. Nowhere is his gratitude more visible to the mother of sciences, geography, than in the creation of the Bhoovigyan Vikas Foundation

For geography

At age 71 when most are tired and ready to retire, Sundaram took upon his frail shoulders the inception and promotion of the Bhoovigyan Vikas Foundation. It was in the aftermath of the 21st Indian Geography Congress held at Nagpur on the 2–4 January 2000, that the need was felt to bring the diverse earth sciences including geography under one umbrella organization. Whereas over 400 delegates discussed and pondered on the ways that earth sciences could be given their due recognition Sundaram announced his decision to start the Foundation.

Sundaram worked swiftly. In August of the very same year the society was registered. A website was launched by December. A logo depicting the

Earth and man spells out the philosophy of bringing together all the disciplines of earth sciences to nurture the earth. The programmes set out are also meaningful. The Foundation would carry forward the mission to care for the earth by creating awareness by creating a museum of Earth and celebrating significant events like Earth Day. The Foundation was launched in Delhi on 21 April 2001 with celebrations marking the first Earth Day in which more than 5,000 school children participated and took pledges as Earth trustees.

In just five years Bhoovigyan Vikas Foundation had enrolled over 200 members. In the brief half a decade since its inception, it could boast a list of 26 publications including 14 plenary lectures, three newsletters, five seminar proceedings and six books. Three research projects were completed under its banner while a dozen others were pipelined. It had held three major international seminars. Sundaram truly set up the Bhoovigyan Vikas Foundation seminars with personal zeal. Confirming Sundaram's ideological stance is the title and theme of all three seminars which revolve around development, sustainability and the rural. True to its commitment, in a short time span the Foundation honoured 24 earth scientists with awards ranging from leadership to work of lifetime excellence.

Anyone who has worked at establishing a society or a non-government organization knows the endless committees, letters and meetings that must precede its birth. What is more, the Foundation branched out to be inclusive. Among its members were scientists belonging to 24 different branches associated with the Earth; these could well be geographers, geologists, agriculturalists, biotechnologists, hydrologists and those engaged in remote sensing and other information sciences. The seminars netted a high attendance with participants from almost every state in India along with half a dozen from other developing countries. A similar diversity of representation is apparent in the recipients of awards and applicants for research projects. The Foundation has, as its patrons, many eminent personalities of the country. Covered by national dailies like *Pioneer*, *Indian Express*, *Economic Times* and *Hindu*, the Foundation also drew the accolades from the likes of chief ministers, governors and other high-placed dignitaries.

For Sundaram, championing the Foundation's cause turned into a personal mission. He not only staked his old age savings and used his home as an office but also personally travelled around the country and visited geography departments and other institutions to deliver speeches and sermons to raise awareness and concern for the earth sciences and hence the Earth. In a near desperate search for a donor who would be willing to fund the dream of an earth care museum and institute, Sundaram personally wrote hundreds of letters and sent repeated requests to philanthropists, private companies, international agencies, ministries and acquaintances. While all his friends and associates were complementary, not one single individual or organization offered a cheque. Ironical as it may seem a society for earth science remains today without a piece of the "Earth"!

I have had the opportunity to examine the workings of India's leading geographical associations. It is with complete conviction that I can claim that what Bhoovigyan Vikas Foundation achieved in a short five years is what many others could not in 25 years. Perhaps one needs to be aware that India's top geographical associations, though nourished within the ambit of India's leading universities, have not been able to build even a small room to run an office. None of the geographical associations in India have taken the bold step of setting an alliance with other sectors of earth sciences. Moreover, none of the geographical associations have even perceived of floating the Bhoogol Ratna Awards or launching an Earth Day celebration. But at times comparisons are small consolations.

When Sundaram handed over the office of chairman of Bhoovigyan Vikas Foundation he carried a feeling of deep remorse. The Foundation had not been able to garner the required funds to set up an institute; it had failed to acquire a small piece of land nor was it successful in establishing the Earth Science Institute or the Earth Care Museum. Did the Foundation fail? While it is too premature to answer this question, a deeper thought would be to inquire: did the Bhoovigyan Vikas Foundation flounder or was it the languishing status of geography and more so, geographers, which are responsible for the state of affairs? After all the fire which led to the birth of the Foundation was the dismal recognition that geography attracted in India.

Nepal, Bangladesh, Ghana, Uganda and geography in India, all echo the reality that maladies of underdevelopment are difficult if not impossible to overcome. The Bhoovigyan Vikas Foundation was an earnest attempt to change this. Sundaram did not miss any opportunity to fight for what he thought was needed; in this regard his example of setting up the Foundation will probably be his greatest legacy. One of his last campaigns was not waged for himself but rather for the survival of geography in India. The inability of the Foundation to gain institutional recognition does not bode well for the future of geography or, for that matter, earth sciences at large.

During the Bhoovigyan Vikas Foundation years Sundaram was a driven man, self-driven. He burned the candle both ends, but that does not give you two candles. The outcome: his health gave way. In these circumstances he had little choice but to resign the chairmanship in April 2009. Today he might occasionally complain of his ailments, but is nevertheless full on as always…suggesting plans for geography and development ….

I cannot but end with an anecdote of where and how I fit into the scheme of things for this write-up. If memory serves I first met Sundaram in 1995 in connection with a seminar I was organizing on careers in geography. For this I was in search of a geographer–planner. I had known of Sundaram's works but never had the opportunity to meet him.

Since this first interlude I have been in constant touch with Sundaram. He has enriched my life in many ways but there is one I will always cherish. The first really encouraging thing I ever heard from the wider academic community was from Sundaram. I had written my first work on Indian geography called *Geography in India: A Future with a Difference*. When I posted it to Sundaram a prompt reply full of appreciation followed. As I was experiencing the early moments of academic maturing it was the Bhoovigyan Vikas Foundation which bestowed the Leadership Award on me. I was delighted. It was the first time I realized that I was doing something that might be appreciated by people for reasons other than their direct institutional or personal relationship.

More than the academic feedback though, what inspired me to do this write-up is Sundaram, the person. His self-discipline, dedication to deliver, and genuineness are qualities so touching that they move the soul to want to know more and more of him…

Write about a geographer more seasoned in age and far more accomplished gets no easier as the years roll by. I first wrote about the life of a geographer, my supervisor, when I was aged 34 years. Since then I have written about more than two dozen other geographers of India. Today I am 50. I write from my room in the Department of Geography, at the Delhi School of Economics and look out longingly at the *shisham* outside my window. Under the canopy of this large deciduous tree I often perch on my chair and sit to read. As I set my pen down and glance again at the tree I ask myself whether I have done these pages justice; is this a fitting tribute for an 80th birthday celebration of one who is not just a master geographer–planner but carries the aura of a grandmaster?

I gifted this write-up unpublished to him on his 80th birthday. Sundaram died in 2013.

This is being published in 2021. Sometimes it is hard to explain the workings of time.

Acknowledgements

While assisting me on the write-up of K.V. Sundaram, my students Yukti, Jenifa, Anuradha and Punam provided valuable time to see it through its completion. Girja, Sundaram's wife of warm-hearted hospitality allowed me to stretch my interviews over endless rounds of South Indian coffee.

Notes

1 To identify the discipline of the contributors in Sundaram's works, the following edited books were selected:
   i. Sundaram, K.V. (ed.). 1977. *Rural Area Development, Perspectives and Approaches*. New Delhi: Sterling Publishers.
   ii. Sundaram, K.V. (ed.). 1985. *Geography and Planning: Essays in Honour of Prof. V.L.S. Prakasa Rao*. New Delhi: Concept Publishing House.
   iii. Sundaram, K.V. 1997. *Decentralized Multilevel Planning: Principle and Practice, Asian and African Experience*. New Delhi: Concept Publishing House.

    iv. Sundaram, K.V. and Nangia, S. (eds.). 1985. *Contributions to Indian Geography, Vol. VI: Population Geography.* New Delhi: Heritage Publishers.

    v. Sundaram, K.V., Ramesh, A. and Tiwari, P. (eds.). 1989. *Regional Planning and Development: Essays on Space, Society and Development in Honour of Prof. R.P. Misra, Volume I.* New Delhi: Heritage Publishers.

    vi. Sundaram, K.V., Ramesh, A. and Tiwari, P. (eds.). 1990. *Regional Planning and Development: Essays on Space, Society and Development in Honour of Prof. R.P. Misra, Volume II.* New Delhi: Heritage Publishers.

    vii. Sundaram, K.V. 1983. *Geography of Underdevelopment: The Spatial Dynamics of Underdevelopment.* New Delhi: Concept Publishing Company.

    viii. Sundaram, K.V. 1977. *Urban and Regional Planning in India.* New Delhi: Vikas Publishing House.

Bibliography

Bacon, Francis 1561–1626. *On Studies*, in *The Essays*; or, *Councils, civil and moral.*

Geddes, Patrick 1968. *Cities of Evolution* London: Benn.

Green, Harlod Louise 1955. "Hinterland Boundaries of New York City and Boston in Southern New England." *Economic Geography*, 31 (4): 283–300.

India News and Feature Alliance. 1969. *India – Who's Who.* New Delhi.

Kapur, Anu 1998. *Geography in India: A Future with a Difference.* New Delhi: Allied Publishers.

Kapur, Anu 2002a. "Kavasseery Vanchi Sundaram: A profile." In *Voice of Concern*, 377–379. New Delhi: Concept Publishing Company.

Kapur, Anu 2002b. "Vidhiparti Lova Surya Prakasa Rao: A profile." In *Voice of Concern*, 257–259. New Delhi: Concept Publishing Company.

Kapur, Anu 2002c. "Rameshwar Prasad Misra: A profile." In *Voice of Concern*, 347–349. New Delhi: Concept Publishing Company.

Khan, Akhtar Hameed 1984. "Comilla Projects and their Relevance." *Pakistan Journal of Public Administration* 23 (1 and 2): 18–38.

Mishra, Rameshwar Prasad (ed.) 1990. "Dr K.V. Sundaram: A profile." In *District Planning: A Handbook*, 311–324 New Delhi: Concept Publishing Company.

Mukherjee, A. 1990a. "Studies in Multilevel Planning: I Researches in Decentralisation *With special Reference to District.*" *Planning in India*. District Planning Unit, Lal Bahadur Shastri, National Academy of Administration, Government of India, Mussoorie.

Mukherjee, A. 1990b. *Foundation of Decentralisation with Special Reference to District Planning in India*, Volume 2. New Delhi: Heritage Publishers.

Myrdal, Gunnar 1957. *Economic Theory and Underdeveloped Regions.* London: Methuen.

Nair, Kusum 1987. *Blossoms in the Dust: The Human Element in Indian Development.* University of Chicago Press.

Perroux, François 1950. "Economic Space: Theory and Applications." *Quarterly Journal of Economics* 64(1): 89–104.

Planning Commission. 1972. *Summary Record of Twenty Eighth Meeting of the National Development Council*, May 30 and 31. New Delhi: Government of India.

Planning Commission. 1989. *Training for District Planning – Report of the Study Group*. New Delhi: Government of India.

Sauer, C.O. 1956. "The Education of a Geographer." *Annals of the Association of American Geographers* 46 (3): 287–299.

Schuster, G. 1950. *Report of the Committee on Qualifications of Planners*, (Cmd.8059), Her Majesty's Stationery Office (HMSO), London: Ministry of Town and Country Planning.

Sharma, R. 1998. *Structural and Functional Analysis of Standard Urban Areas in India*. Unpublished PhD thesis. Department of Geography, Delhi School of Economics, University of Delhi, New Delhi.

Sundaram, K.V. 1985 "Prakasa Rao, the Planner." In *Geography and Planning: Essays in Honour of Prof. V.L.S. Prakasa Rao*, edited by K.V. Sundaram, 32–36. New Delhi: Concept Publishing House.

Sundaram, K.V. 1989. "Prof. R.P. Misra: A Review and Assessment of his Contributions." In *Regional Planning and Development: Essays on Space, Society and Development in Honour of Prof. R.P. Misra, Volume I*, edited by K.V. Sundaram, A. Ramesh P. Tiwari, 13–26. New Delhi: Heritage Publishers.

Ullman, Edward 1941. "A Theory of Location for Cities." *American Journal of Sociology*, 46 (6): 858–863.

Wise, M.J. 1996. "Bertram Hughes Farmer, 1916–1996." *Transactions of the Institute of British Geographers* 21 (4): 699–703.

World Bank. 1990. *World Development Report 1990: Poverty*. New York: Oxford University Press.

Part III

LifeContours

6 The life and works of 22 geographers in India

Gurdev Singh Gosal

In the class of learned men Gosal can be bracketed as a university man. Born on 1 April 1927, he began teaching at Government College, Ludhiana, Punjab in 1950 and joined Panjab University in 1959. With a good 70 per cent of his life spent in teaching and research, the lure for an educational environment beckons him, even at the age of 74, to vigorously getting down to completing his book *History of Geographic Thought*.

Gosal enjoyed the university and this enjoyment was requited, with the university quick to give accolades. As if standing on a high-speed elevator – he was awarded a doctorate in three years (1953–1956), appointed lecturer at the age of 23, became chairperson of the department at 34, professor at 38, elected member of the university senate at 39, dean for student welfare at 40, dean of university instruction at 53, dean of college development council at 59, and many other university ranks. Gosal, however, was not driven by the automatic levers dispensed by position and power. That he thrice refused the position of vice chancellor proves that he could exercise conscious choice to switch off the temptation to land at this highest rung in the hierarchy of a university.

Privileges of family, social contacts, and burning ambitions are often quoted as crucial elements which lend force to the gradient of a career graph. None of these factors operated in Gosal's case. Belonging to a cheerful family of farmers, in the village of Goslan, in district Ropar, Punjab, the settlement gave him his surname and his father the spirit to always be first. His family was neither literate nor affluent. A self-made man, Gosal abhorred the snares of socio-political networking. Instilled more with determination than ambition it is Gosal's grand clarity of thought that quickly cut his professional journey.

While still young he decided that teaching and not farming was to be his vocation. Backed by a first-class first he was appointed as a college lecturer straight after post-graduation. Quick to discover that a sound bedrock of learning is an essential prerequisite to impart knowledge, he set out to search of a reputable institution. By the early 1950s the US was a more promising destination for geography and had begun to supersede the UK. Weighing his options, it is Gosal's ability for alert appraisal which won him the Whitbeck Fellowship in 1953 and took him to the University of Wisconsin in Madison. In the 1950s, Hartshorne, Robinson, Trewartha and Olmsted were among the galaxy of renowned faculty of geography at Madison. Gosal obviously knew how to make a near perfect choice for himself. An unclouded hindsight signalled him towards the study of population geography under the grandmaster Trewartha; wrench it until it is all meat was Trewartha's steely philosophy towards research. All of Gosal's 50 maps in the dissertation had to meet the gruelling scrutiny of Robinson, the famous author of the book *Elements of Cartography*. Armed with a strong intellectual muscle, mincing without wincing, the meat of Gosal's dissertation titled, "Population Geography of India," found approval in just 24 months. If Wisconsin was a golden period of Gosal's apprenticeship, building a department of excellence in geography at Panjab University with a thrust on population geography in India was his undeterred dream.

Gosal's commitment to his specialization had multiple ramifications. Motivating a handful of students to work on different aspects of population and recruiting them at an early stage in the department may seem like inbreeding, but it ensured the perpetuation of his objective. Founding an Association of Population Geographers of India and launching its journal, *Population Geography* in 1979, were landmarks not only in the country but on the international circuit too.

When the newly established mapping division in Census of India in 1962 was searching expertise to prepare its Census Atlas of Punjab, Gosal was an obvious choice. He not only became the chief architect and executor for the project but the work carries the stamp of his grand expertise. That 30 of his 37 research papers and 11 of the 12 doctorates he supervised revolved around the axis of population is proof of Gosal's devotion to the study of population geography. To stay abreast with research, Gosal participated in the International Geography Union Commission on Population Geography and International Symposiums at Pennsylvania State University 1967, London 1969, Canada 1972, Australia 1988 and Germany 1989. The faculty, association, journal, research papers, seminars, symposiums, and works with the Census of India were strands Gosal braided together to strengthen the Department of Geography at Panjab University, Chandigarh.

Those were the days when department chairpersons were not rotated like a game of musical chairs. Gosal's unchallenged headship for two decades from 1961 to 1981, could and did pipe the music closest to his heart: population geography. It is, however, the quality rather than the sheer volume that put the Department of Geography, Chandigarh into the limelight. Research assistance, grants, and projects flowed to the department from diverse quarters such as the Ministry of Defence and the Indian Council of Social Science Research and, in later years, the prestigious status of Special Assistance Programme from the University Grants Commission was accrued from these years of work. It is no surprise that the first Indian Geography Congress of the National Association of Geographers, India 1980 was held at the Panjab University Department of Geography, Chandigarh and its first president was none other than Gosal.

Population geography was not his exclusive domain of interest. Even without personal contact, Hartshorne's influence on geographers is omnipresent and Gosal had the good fortune to be his student. Stimulating lectures by Hartshorne on the study of geographical ideas and thoughts cast an indelible impression on the young Gosal. When the load of administration bit into his time, Gosal dropped teaching population geography, but continued to lecture on philosophical aspects of the discipline. His legacy of six generations of students at the department certify that his lucid grasp over the subject caught their unblinking attention.

Scholarship, poise and dignity are character traits which can be collected in the quiet chambers of reading, research and writing. Unfortunately, they cannot always be confined. The erudite Gosal was drawn into the echelons of the University of Panjab. Elected successively as member of the senate and syndicate while he tried to balance academics with administration, he claimed with wit that, "his black beard quickly turned white." The responsibilities did steal, perhaps a little too early, the solitude for reflective thinking in furthering geography. It must be remembered, however, that universities are kept alive not merely on the shoulders of researchers and teachers, and it is only when the structure of the entire system is healthy that scholarship can be upheld. Further, it is not mere books and research, but the faceted experiences of life that channel wisdom. Gosal enjoyed the university in a wide spectrum of ways. He epitomizes the best in a university man. (1927–2014)

Hari Prasanna Das

In 1952 when Das boarded *Majola*, a warship converted to a passenger ship, at Bombay, and set out for the London School of Economics to acquire a doctorate he had two clear missions: to study the geography of Assam and spread geography in Assam. Completing his doctorate under the supervision of Stamp in a short 21 months Das returned home with a thesis entitled, "Forest Resources of Assam." On the foundations of his first mission, he quickly laid the blocks of other publications: *Problem of Forestry in Assam*, 1958; *Land Use in Assam*, 1959; and *Geography of Assam*, 1970. The additional dozen publications and seven doctorates he supervised all confirm that Assam was the nerve centre of his research interest. Even with limited publications, Das made an unparalleled contribution to the meticulous understanding of the geography of Assam. He also had a second resolution to fulfil – the dissemination of geography as a discipline of study in the state. Serving as chairperson for the Department of Geography, University of Gauhati until his retirement in 1984, he not only inaugurated the Geographical Society of the North-Eastern Hill Region in 1968 and floated the *Journal of North-East India* in 1970 but also unfurled the cause of geography outside the confines of the department. His persistence opened the way for geography in schools and colleges of Assam and provided opportunities which teachers were quick to seize. To propagate the discipline, Das published a four-volume set of *Adarsh Bhoogol* (Model Geography in Assamese), which became a popular text for Standard 7 to 10 classes in schools.

Das knew that to safeguard the study of the discipline the smooth functioning of university administration was critical. He worked with vigour not only on a number of committees but also took on various administrative posts in the university. He served as dean of the science department, rector of the university and was also acting vice chancellor for a term. Responsibilities such as director of the sports board and warden of hostels also fell on his shoulders. As secretary management of colleges for over a decade, Das was associated with the administration of a dozen colleges in Gauhati. With growing involvement in these multifarious activities, Das neglected his personal research but not his pupils. Though conventional in approach the assiduously prepared lectures made Das an esteemed teacher. He firmly held the belief that geographers had to first serve and understand their own environment. Assam was a state which strongly captured his interest. Nostalgically he reiterates that even a short journey outside Assam makes him homesick. More than a provincial zeal, Das was keen to serve the region through the medium of education in the realm of geography.

Das's bonding for the land and people of Assam runs long and deep. It is his family that kindled this patriotic sentiment. Das was born on 1 March 1922 in a village called Kaithalkuchi, about 80 km west of Gauhati. Nalbari was the closest town and it was here at the Gordon High School that Das pursued his basic studies. He graduated from Cotton College and then enrolled at the Department of Geography, University of Calcutta, for his post-graduation in 1946. His father, a famous headmaster in the Gordon High School, had studied geography for his Bachelor of Teaching degree. While Das inherited a passion for geography and teaching from his father, the attachment for the study of Assam was sown when he journeyed through inaccessible tracts of this mountainous state to assess its forest wealth. He recalls that, as a research fellow in the aftermath of a postgraduate degree, the inclination to study forest resources of Assam found a rewarding outlet when the Chief Conservator of Forests, M.C. Jacob, rendered all help and guidance. Extensive fieldwork into the remote interiors of Assam aroused in Das an ardent desire to serve this less developed state of India. Today, despite faltering health, many seek his wise counsel about geographical issues pertaining to the region. And why not? After all, throughout his four decades as a professional geographer, Das remained undeterred from his goal to study the geography of Assam and spread geography in Assam. (1922–2003)

Balsubramiam Arunachalam

If a picture is worth a thousand words, then a map is a picture worth a lot of condensed spatial information. All geographers use maps for research, some design and compile maps to build an atlas while some teach the science of map-making. Arunachalam's relationship with maps goes beyond these mundane uses. He searches for indigenous maps of India, probes the mathematical principles upon which the foundations of these maps were raised and infers historical geography from them. Arunachalam taught geography in colleges and at the University of Bombay for 40 long years. He used the strategic location of the city of his professional placement to further his study of maps. Bombay, India's leading port for maritime trade and naval movement, depends upon the use of nautical charts. Arunachalam thus chose to specialize on maps of maritime navigation. Fishermen, boatmen to the officers, and chief hydrographers of the government of India were all a repository which Arunachalam tapped to solicit knowledge about maps. Akin to an adventurous schoolboy, he traversed from Kutch to the Sundarbans, scouting the 7,000 km coast of India. Sailing on weather ships, cargo vessels, rowing boats, or with coastguards, Arunachalam ventured out into the sea to reach off-beat islands in the Bay of Bengal and the Arabian Sea. One question that troubled Arunachalam as he made these field trips was how, without formal education of latitude and longitude or the use of sextants and compass, the native boatmen designed maps? Pointing towards the stars the natives unfolded the mystery of drawing near perfect maps to Arunachalam. To exhibit their skills the boatmen often used the sand beach as a slate to illustrate the maps. Discussions in the chambers of officers of naval and shipping companies furthered Arunachalam's comparative understanding of map-making. Traditional texts, archives and antique shops selling rare books were additional haunts from where Arunachalam retrieved maps of the past. Such being the passion, it is no surprise that the Department of Geography at the University of Bombay can boast of a collection of some eight to nine thousand maps. The rich variety of cadastral, topographic, naval and historical maps, along with fascinating editions of atlases of all kinds earned the Department of Geography at Bombay a reputation as one of the national reference libraries for maps in India.

Arunachalam is not just a traveller and collector, he is also a researcher. Forty of the 60 papers authored by him orbit around the theme of maps. Three of the four research projects he investigated had the study of maps as their core concern. Both in historical perspective and content, Arunachalam picks a variety of topics for research. Calligraphy in cartography, indigenous traditions of navigation of South India, Jain pilgrim maps, the cartography of the Chola administration all exemplify this vast sweep. Arunachalam's research is not static. His inquiry into medieval traditions of Indian cartography was quickly followed by that on indigenous traditions of Indian cartography. The skills of coding local maps to the interpretation of aerial photographs, imageries and application of geographic information systems are proof of his evolving research. Papers on digital terrain models and on isopleth mapping through computers and geographic information systems are ingredients in his list publications. Prolific in number are his works, yet Arunachalam's professional interest in maps was aroused late. He had spent 20 years in teaching and research, and already passed his 40th birthday when his inquisition was moored to the study of maps.

Born on 1 July 1933 in Calicut, Arunachalam went to school in Cuddalore, a small town 22 km to the south of Pondicherry. With physics, mathematics and geography at intermediate level, geography was not a subject of first preference. Arunachalam wanted to study statistics. Though a second preference his performance in the study of geography was outstanding. Scoring a first-class pass he won the meritorious Senior Thompson Prize of Presidency College in 1953. Despite academic distinctions Arunachalam's formal education terminated with a master's degree. Arunachalam never took a doctorate, never went abroad to study and he never took a teaching assignment at a foreign university. In this sense Arunachalam is a self-learner. A genetic disposition from his father, who was a middle school teacher, could partly explain his scholastic aptitude but Arunachalam had no intellectual mentor or supervisor. At the young age of around 20 he was appointed lecturer at the Ruparel College in 1953 and shifted to the Parle Commerce College in 1960. When the Department of Geography was started at the University of Bombay in 1969, Arunachalam soon found a placement in it a year later.

In this first 20 years of teaching, Arunachalam's research interests hovered around the regional geography of Maharashtra and geomorphology. A keen desire to disseminate the importance of geography motivated him to write a series of school textbooks and regional texts on Maharashtra. His interest in geomorphology is displayed in his research publications on *Coastal Features in the Vicinity of Ratnagiri*, *Shifting Water Divide in the Gatpuri – Trambak Ghats*, and the *Coastal Changes in the Kochi Embayment*. Using satellite imagery and topographical maps along with primary field observations, Arunachalam's publications even in this arena carry the hallmark of quality. How then did Arunachalam shift to the study of maps?

A historian working on an assignment brought a sheaf of 300-year-old maps to Arunachalam to decode in 1976–1977. The task ignited his imagination and captured his interest. To crystallize his research Arunachalam displayed remarkable dexterity. He was one of the founder members of the Indian National Cartographic Association and became its president in 1990. An active interest in the Society for Indian Ocean Studies and Maritime History Society was an additional avenue that Arunachalam utilized to forge ahead in his study of maps.

Today at the age of 68, despite a coronary problem he is pursuing two ambitious projects – first to establish a Maritime Museum in Bombay, and second to compile his life work in a book titled *Native Indian Cartographic Traditions*.

While most researchers are engaged in conventional themes of land use, urban studies, and regional planning, Arunachalam has indeed carved a unique picture. This is why among Indian geographers, Arunachalam's name with maps is inseparable. Whereas geographers use maps as tool for research, Arunachalam researches maps. (1933–2014)

Moonis Raza

On a summer evening of 18 July 1994, in the city of Boston, USA, a cardiac arrest bundled Moonis Raza to the carpeted floor. He died at 5.00 PM with no one at his side. Moonis was buried at the city graveyard, where a small group of family members attended his funeral. There was no geographer to lay the wreath. Moonis was not a non-entity but one of the most charismatic geographers India has ever produced. His flair for language, captivating extempore speeches, powers to organize and generate brilliant ideas see few parallels in the academic landscape of Indian geography. Had Moonis died on native Indian soil, dignitaries, politicians, heads of leading institutes, colleagues, friends and students from such diverse disciplines as geography, history, economics, sociology, demography, regional science and education would have thronged in numbers large enough to create a stampede. Such was the personality of Moonis that few could resist him. He had his way with most people in most situations.

Cutting a career in a varied terrain, Moonis notched landmarks which hardly any other geographer has been able to surpass. No geographer in India has taken the prestigious Shiromani Award, nor has anyone been the principal of an engineering college, or the chairperson of the Indian Council of Social Science Research. No one has enjoyed a second nomination as president of the National Association of Geographers, India and to date, no geographer in India has supervised a dozen doctorates without a doctorate degree to his own credit. He was one of the few geographers in India who grasped the patterns of disparity on a pan-India scale. No other geographer in India has enjoyed a loyal coterie of followers – who not only laboured and worked in accordance with his ideas, but whose research even today ripples like concentric circles around the creative ideas set in motion by Moonis. Many geographers in India have attained recognition but none were charismatic in the Moonis way.

Moonis was not the only gifted child of his parents. Rahi Masoom Raza, the man famous as the scriptwriter for the dialogues in *Mahabharat*, a popular television series in India, was his elder brother. Born on 2 February 1925 in Ghazipur, a district headquarters in eastern Uttar Pradesh, Moonis lost his mother at a young age. With a father who was a prosperous advocate in the district court Moonis and his brothers and sisters were left to fend for themselves. A common bond among the siblings was a stock of some of the finest English and Urdu classics which their father had collected over the years. Few would deny that it is the prolific reading in our young days which fills a reservoir of words that spring to life when we speak and write in years to come.

Hailing from such a family, Moonis set out to do his graduation from the Christian College, Allahabad. For post-graduation it was the Department of Geography at Aligarh Muslim University which became a carefully considered decision. From 1945 to 1947 Aligarh played a critical role in Moonis's life. The raging student politics on the campus, rather than the Department of Geography, shaped and sharpened the analytical abilities of the 20-year-old Moonis. At the heel of India's independence, those were times when political organizations and rallies ran rampant. Moonis became the student leader of the Communist Party of India at Aligarh. His commitment ran deep and severing ties with the family, the labour camps at Kanpur became his refuge for many years. After three long years when Moonis returned home to Aligarh in 1950, his alma mater offered him a position as lecturer in 1950. A malignant tuberculosis and threat of a police arrest for his questionable political leanings forced Moonis to resurface and seek a haven in academics.

Home is where the care is and the sick lad needed nursing. Nothing would cure him and Moonis had to be rushed to Czechoslovakia for a lectimoy lipectomy. This resulted in a marked reduction in the capacity of his lungs and affected his breathing. Alongside this physiological condition, the politically active years had left a deep intellectual and emotional impression on Moonis. The killer instinct to survive, the intense ability for analytical thinking, the adept power to assign work and a heartfelt care for the downtrodden were characteristics which gave Moonis an immediate edge to fascinate one and all. He used his charm to hold office and inspire officers. But he was no superficial charmer – a burning passion to throw light and pay attention to problems of social and spatial disparities were his academic objectives. He recognized that the shortest path to reach this goal was not through geography alone. Moonis was drawn to the field of education at large. Though his interest in regional development was sustained, education as a field of study and work came to distinctly dominate. Here is the audit: of the 15 research projects that Moonis undertook, 40 per cent were on education and 14 per cent on regional development; of the 19 papers presented at international seminars an almost equal share of 16 per cent belong to education and regional development; of the 31 academic addresses delivered as lectures on special invitations 13 were on education and only two on regional development. Forty-six pieces of writing can be counted by Moonis as author, co-author and editor of books. Among these though 20 per cent belong to regional development and 13 per cent to education, the largest, an 11-volume set, is on the theme of higher education. In the domain of his 36 research papers published in various journals, 13 are on education and only six on regional development. Of the eight different administrative positions enjoyed by Moonis half have been within a government institute of education. But these were not his only pursuits – urbanization, medieval and colonial history, ecological issues and resource geography embraced his interest. A theme running common to most of his works was the search of unity in diversity. In fact it is in synthesis that lay Moonis's greatest talent. Artfully knitting past with future, tradition with modern, micro with macro, spatial development with socio-economic parameters, scientific with poetic language, he could integrate all this well.

Moonis above all was a people's person. Granting favours, deciding honours, discussing ideas and debating issues; in all hospitality and at great ease Moonis was perpetually surrounded by students or colleagues. He had a renewable spring of ideas to air and share. He thus enrolled many into his fold of creative thinking, initiating numerous joint works and accomplishing many a mission. More than heading offices and pushing paper it is the brilliance of Moonis's words and ideas that set him in a class apart.

It is indeed a twist of irony that a person so full of life should remain in stone silence for 22 painful months in the Intensive Care Unit of the All India Institute of Medical Sciences, New Delhi. A medical decision for treatment in the USA distanced him from his people. The misery of a half lung, uncared for in the zest to lead a full life, had taken its toll. Though the body gives way, a charismatic soul lives on forever. (1925–1994)

Nalagatla Bala Krishna Reddy

Reddy was born on 1 July 1929 on port city on the Tammeleru river in Andhra Pradesh called Eluru.

When Reddy's father admitted him at the age of five into the Ramakrishna High School little was he aware that one day his son would be the chairman of the Andhra Pradesh State Council of Higher Education. After all no one in Reddy's father's or forefathers' generation had been to school. Belonging to a family of semi-landlords, Reddy recalls his parents may not have been literate but a liberal intellectual environment did exist within the home. Heated debates and lively discussions on the Indian National Movement for example were a topic of common conversation. The milieu charged Reddy towards active politics.

In his days of youth, Reddy was not an armchair communist; for him organizing and participating in meets and demonstrations was a common practice. He was the general secretary of the City Students Union at Karnatak University in 1948–1949, of the Government Training College Teachers Association in Nellore in 1956–1957 and of the Andhra Association at Banaras Hindu University in 1958–1959. It is the attention Reddy devoted to spearhead his political commitment which drafted him into study geography. Not that active politics and geography have much in common but in the case of Reddy there is a relationship. Reddy wanted to pursue medicine. For his bachelor's degree he had zoology, physics and chemistry as his courses of study. But the hours spent in political meets were a distraction which took precious time away from preparing for medicine and thus forced him to take up geography. Politics perhaps had come his way only to veer him towards an academic course. As soon as he completed a master's in geography, his interest in politics waned and he set his mind to books.

Reddy's intellectual ability is apparent on many fronts. Abstaining from the bandwagon of politics his grades jumped from second to a first-class first in his master's degree. Marxian ideology had influenced his youth but did not colour his research. A scientific temperament reinforced by his training in the science subjects instilled a marked objectivity in his research. Though Reddy had a greater interest in physical geography he was aware of his limitations and opted for a doctorate on, "Urbanisation in the Krishna and Godavari Deltas." His initial research avoided physical geography; his techniques of analysis even with the human parameters were not lay descriptions. With publications based on refinement of nearest neighbour and reflexive neighbour analysis, delimitation of polynuclei urban zones and moment analysis he contributed to the methodological facet of urban geography. Set on a hard rock of quantitative analysis, quality is the hallmark of Reddy's publications.

The attraction for physical geography led him to undertake a certificate course in geomorphology and photogeology in 1978, a decade after his doctorate. Thereafter his research spanned both the physical and human aspects of the discipline. Nearly one-third of Reddy's 60 research papers, four of his eight research projects and seven of the 17 students he supervised for a doctorate worked on physical geography. Reddy was not just a researcher; his ability to impart knowledge was duly acknowledged by the government of Andhra Pradesh as they conferred on him the honour of Best Teacher Award in 1984. Here Reddy's bachelor's and master's degrees in education were additional assets. Endowed with prerequisite skills after an initial period of unsettlement, with a year at the Government Arts College at Srikakulam, 12 years at the Karnatak Science College, and a year at the Department of Geography at Karnatak University, Dharwad he finally settled to a devoted tenure as founder and chairperson of the Department of Geography at the Sri Venkateswara University, Tirupati, from 1972 to 1989.

To live life as a recluse around research and teaching was not the way of Reddy. Seeking a wider interaction he not only took frequent sojourns to attend a dozen conferences in India and half a dozen abroad but also sought active participation in the various geographical associations in India. He was the president of the National Association of Geographers, India as well as the Indian Geographical Society, Madras; the Institute of Indian Geographers, Pune; and Indian Council of Geographers, Bhubaneshwar. Reddy's wise counsel and friendly cheer to dissolve vexing situations also won him a place in numerous administrative posts at the university. He was a member of the senate and academic council and one of the trustees on the management board of Sri Venkateswara University, Tirupati. Despite his prowess, Reddy could not become the vice chancellor of a university but, he says, a better proposition came his way. On his retirement in 1984 he was appointed as chairman of the Andhra Pradesh State Council of Higher Education, a position where he liaised and coordinated the activities of vice chancellors of ten universities in Andhra Pradesh.

At the age of 72 Reddy sat on the veranda of his house compiling his book *Patterns of Political and Cultural Antecedents of India: A Geopolitical Analysis* and looking out at the road which leads to the famous temple of Sri Venkateswara on the Thirumala hillock of Tirupati, one fact is clear that from Eluru, Reddy has indeed come a long way. (1929–2006)

Mohammad Shafi

"Mecca of Indian Geography," is what the famous British professor of the land use survey, Professor L.D. Stamp, called the Department of Geography at Aligarh. A considerable share of sustaining this compliment accrues to Shafi.

Shafi is not the founder of the Department of Geography at Aligarh established in 1924, he joined the department as faculty 24 years later. Shafi belongs to the eighth batch of postgraduate students who passed out of the department. The Curzon Geographical Society and the journal *The Geographer* existed 22 years before Shafi's first contribution in 1948. The long-standing department, name of the society and its journal, *The Geographer* existed years before Shafi made his presence felt.

Carrying the hallowed glory and traditions of excellence of the department forward, it is the recognition of Shafi's research as a doctoral student under the tutelage of Stamp in 1956 which makes him a critical partner in the compliment. Acknowledgement by scholars of repute, especially those belonging to universities in the developed world, places one immediately in the limelight. In the conservative environ of British universities, Shafi met his supervisor half a dozen times. Yet the inclusion of a sizeable reference from Shafi's doctoral work in the presidential address of Stamp at the International Geographical Union in Rio de Janeiro in 1956 and its publication in the *Geographical Review* in 1957 affirms the perception he had of his student. Despite the frugal meetings, Shafi worked tenaciously and submitted his doctorate in a brief span of 21 months.

A vortex of head-spinning success gravitated towards Shafi when, in 1956, he returned from UK to Aligarh, the city which was to remain his permanent place of work and stay. He was promoted to reader in 1959. He still hadn't reached his fortieth birthday when he was appointed professor and chairperson of the department. His tenure of 22 years as chairperson was the longest that the department ever had. Honoured as professor emeritus on his retirement, he has a record of extended services at the department. In terms of the frequency of attendance at the International Geographical Congress, Shafi hardly has any parallel. He has the rare distinction of being elected as vice president of the International Geographical Union for two terms. With nearly 10 per cent of all the articles in the in-house journal, *The Geographer* and a sizeable number of books published at Aligarh University, his monopoly in publications is outstanding and his success is not confined within the department of geography. Though Shafi was vice chancellor for only a year, he remained the pro-vice-chancellor for 11 long years. He was awarded with the Padma Shri by the government of India on the occasion of Republic Day, 2001. The same year the Bhoovigyan Vikas Foundation, Delhi, conferred its first Bhoogol Ratna on him.

"A teacher should never rest on his oars," is what Shafi professes. A tenacity to meticulously sift pages for facts built Shafi's reservoir of strength to participate in the discipline of geography in India in a variety of ways. From practical examinations to appointments, class lectures to presidential and keynote addresses, seminars to workshops, occasional writings in newspapers to scholarly books, Shafi performed a multiplicity of roles incumbent on a committed academician.

In Shafi's five decades of professional life, the 1980s and 1990s stands out as a block of maximum activities. It is during these two decades that he attended 50 per cent of 16 international conferences, and participated in four of six national seminars. Forty-one per cent of his 48 research papers were published during these decades. No less than 70 per cent of the 20 honours bestowed upon him fall in this time. While his chairmanship of the department reached its end in 1984, the pro-vice-chancellorship of the Aligarh Muslim University and a brief vice chancellorship also fall within this bracket of time. It is not a coincidence that he was also elected as president of the National Association of Geographers, India in 1993.

Academic work, however, cannot be mechanically contained within a defined span of time. Shafi's exalted endurance of perseverance towards research transcends time. While the stamp of land use has its origin at the London School of Economics way back in 1954, a lifetime of devotion to issues on agricultural geography can be ascertained from his works which initially included themes like land use, agricultural efficiency, agricultural productivity and later proceeded to regional imbalances and cost benefit analysis in Indian agriculture. Shafi's perceptive ability to compare and contrast the characteristics of agriculture in India vis-à-vis the West allowed him to challenge not just the feasibility of Stamp's method of land use survey in an Indian context but also question some of the premise of Von Thunen's model. While the latter had asserted that it is the factor of market which dictated changing land use around an urban area, Shafi's conclusion for the situation in India was that it is the intensity of irrigation which has a more dominant control over land use.

Even at the age of 77 he is engrossed in a chain of works. Barely had his book on South Asia hit the bookshelves in 2000, he submitted his manuscript on Central Asia in 2001. In the pipeline stands the draft of his book *Agricultural Geography – Concepts, Principles and Processes*. With an inexhaustible source of energy, from a primary and secondary schooling in the small town of Jaunpur, Shafi proceeded with his undergraduate degree at the University of Allahabad in 1943 and post-graduation from Aligarh in 1947. Setting sail from Bombay, on Balova, a Polish ship to earn a doctorate at the University of London, Shafi returned home with dreams of hard work and success.

From Jaunpur to London, from lecturer to vice chancellor, from medals in school to the Padma Shri, Shafi has indeed come a long way. He claims, "he does not chase offices but offices chase him." The pursuit continues, for pilgrims never cease. (1924–2007)

Chita Ranjan Pathak

In the drought prone habitat of the Bankura district of West Bengal is a small village called Sonardanga – which literally means dip of gold. To a family of landlords a child was born on 13 July 1935 who they named Chita Ranjan Pathak. The village was remote and even within a radius of 15 km there was no school. Pathak's mother had never been to school and his father had barely reached fourth class. Not only was the level of literacy low, but an extravagant lifestyle had pauperized the wealth of this family of landlords. Yet Pathak's life has been a dip into gold in more than one way.

With a graduation from the prestigious Presidency College, Calcutta, a doctorate from the University of North Carolina, USA, summer courses at the University of California, Berkeley and a post-doctorate at the School of Design, Harvard University in Cambridge, Massachusetts, Pathak's education was a melange within a crucible of the most superior mettle. Purity was added to the lustre not just with the hallmark of these illustrious institutes but from the distinguished teachers who tutored Pathak.

N.K. Bose, the eminent social scientist, lectured on cultural anthropology and the art of field observation at Presidency College. To launch his book *India and Pakistan: A General and Regional Geography*, Spate visited the Department of Geography at the University of Calcutta, where Pathak recalls that his lectures and discussions on regional aspects were most stimulating. Bunge and Haggerstrand took courses at North Carolina in theoretical geography while Marbout and Isaard led the summer course at Berkeley in regional science. At Harvard, Berry was at the helm of affairs in urban geography. Dazzling intellectual energy thus reached Pathak. Reputed institutes and the best of teachers do not necessarily produce the best students. The magnet to learn, after all, lies within the field of the learner. Pathak left his rivals pale with an outstanding academic record at every step of his learning.

Securing first prize for the postgraduate examination, he wrote his doctorate in less than a year and was granted a waiver of 90 credits for the course work at the University of Carolina. His ample knowledge of French and German earned him not just high scores in languages but the fact that he translated Christaller's Central Place Theory from German to English with the help of a friend, denotes that texts in original language provided him with deeper insight for study and review. Pathak perhaps is one among the few geographers in India who has a track record of ten years of scholarships and fellowships. From the Fulbright Fellowship, Smith Foundation and National Science Foundation scholarships, to the Fulbright Alumni Research Grant Senior Fulbright Fellowship and a British Council visitorship, Pathak's treasury received remittances from various sources.

The specialization in regional science which he set out to seek was sterling in nature. For Pathak simplistic and descriptive studies of geographical phenomena were not a challenge. With mathematics, economics and geography as courses of study at an intermediate level and mathematics, Sanskrit and geography at graduate level followed by a specialization in cartography at the postgraduate level, Pathak's aptitude clearly nestled within a quantitative field of inquiry. He wanted to apply geographical concepts to solve problems of the real world. Even in his doctorate on, "Urban Pattern in the Piedmont Region of Raleigh in North Carolina," correlation and pattern analysis formed the core tools for analysis. Skills with the nearest neighbour analysis, simulation studies, gravity models and network analysis drew Pathak towards the newly emerging hybrid discipline of regional science.

Isaard, an economist at Harvard, considered neoclassical economics as a wonderland of no spatial dimensions. It was in the mid-1950s that Isaard founded the discipline of regional science. Pathak was on the rolls of various American universities as a doctorate and post-doctorate student for an extended period from 1961 to 1971. The quantitative and regional science movement captured the attention of the 26-year-old Pathak and became his fixation for a lifetime. In 1967 he founded the Regional Science Association and launched the *Journal of Regional Science* in India. The survival and renown of this journal owes allegiance to Pathak's lifetime devotion. Nearly 60 per cent of his books, half of his research articles, and 30 among a total of 45 papers presented by Pathak rotate around the theme of regional science. A special emphasis is on urban studies, industrial geography and regional planning. He laments that the demise of the quantitative school resulted not so much from the radicalist and humanist schools but the fact that most geographers in India are limited in their ability to handle the advanced techniques of spatial analysis. Pathak's intellectual acumen and specialization had currency value in India.

When the Indian Institute of Technology, raised in an old British prison at Kharagpur, started a course in regional planning, Pathak was recruited as its first lecturer in 1964. When the Centre for the Study of Regional Development was inaugurated at Jawaharlal Nehru University, New Delhi in 1971, Pathak was the first lecturer to be appointed. The sharp intellect of Pathak quickly perceived that the Indian Institute of Technology, an institute of national repute, would be a better option for his creative work. After a stay of five years at the Centre for the Study of Regional Development, he went back to Kharagpur never to uproot himself again. Invitations to a score of teaching assignments from foreign universities and a dozen seminars took him away for six years, but these were intermittent visits; for Pathak Kharagpur was the permanent base. In his 29 years at the Indian Institute of Technology he not only supervised ten doctorates but also attracted numerous projects on district planning, rural industrialization, environmental impact assessment and locational analysis of ports such as Haldia and Digha. Today private consultancies for remote sensing and geographic information systems engage Pathak in numerous ways.

In his lifetime he may not have been crowned with political or administrative posts, or earned pots of wealth, but he surely possesses the golden key of knowledge. In a way he does personify Sonardanga, the village from which he hails. (1935–)

Kashi Nath Singh

If the postman had delivered the admission ticket for the undergraduate course in journalism at the Banaras Hindu University on the stipulated date, today Singh would have been a freelance writer. Teaching was second in Singh's career option list. If, in his graduate days, the geography teachers had not evoked his interest in the subject Singh today would have been an economist. Geography was second in his list of favourite disciplines. Despite a teaching career, the urge to write, with some works hinged on economics, has been an inherent feature of Singh's life. His voracious urge to write is reflected in the stock of 21 books and 90 research papers and over a dozen articles in newspapers. His first textbook *Arthik Bhoogol ke Mool Tattva* (Fundamentals of Economic Geography) carries economics at its core. Not only does Singh have a volume of publications to his credit but many among them attracted rave reviews from both international and national audiences. Here are some randomly picked samples: Spate and Learmonth in their book *India and Pakistan – A General and Regional Geography* refer to his, "unusually vigorous commentary on urban land use to Singh's paper on 'Morphology of the twin township of Dehri – Dalmianagar'", which he wrote the year after his post-graduation in 1957. Another landmark was the research on the, "Territorial Basis of Town and Village Settlements in Eastern Uttar Pradesh." Spencer, the editor of the prestigious *Annals of the Association of American Geographers* journal, judged that the paper is so good that I would like to make it the lead article in the June 1968 issue. Farmer, the Director at the Centre of South Asian Studies, 1975, Cambridge affirms – "the referees appointed by the Institute of British Geographers for the paper titled 'India's Rural Development Strategy and Emerging Patterns of Rural and Urban Relations' have reported most favorably." Back home a review of the *Survey of Research in Geography, 1969–1972*, in the Sunday Magazine of the *Indian Express*, 1979, notes, "the book is like a curate's egg – however, the best piece is the article on Caste by Singh."

How did Singh acquire the prowess to innovative ideas and translate them into scientific writing? Born on 1 January 1932 to a family of small zamindars in Murharia, a village located at the foot of the Kaimur Hills of the Vindhya range in Bihar, Singh's mother never went to school while his father studied up to fifth class. Amidst a family of semi-literates, the essentials for education in the young Singh was instilled by two paternal uncles, one the headmaster of a school and the other a journalist. Singh's consistent devotion towards studies is seen in his mark sheets – a first division and a second position is the standard score in his intermediate graduation and post-graduation. His alma mater was the Banaras Hindu University. In the 1950s and 1960s, the Department of Geography at Banaras Hindu University was popular for studies on urban and rural settlements. As soon as Singh was appointed lecturer in 1957 at his own alma mater, his aim was to strike a departure from traditional research and break into new directions. In his environ where learning was stereotyped, Singh says his drive for innovative thinking came not just from private study of foreign journals and sociological texts but from the opportunities he had to interact with scholars of repute. Submitting his doctorate in 1963 on, "Rural Markets and Rurban Centres in Eastern Uttar Pradesh," a Fulbright scholarship in 1964 took Singh for a year to Rutgers State University in New Jersey, followed by a teaching assignment at the East Stroudsburg University of Pennsylvania. The spin-off from these visits was a close bond with Brush. In his lifetime Singh travelled abroad ten times and visited 23 countries on all six continents, but 1975, he says is a memorable watershed for his road to creative thinking. In this year, Sopher, the geographer from Cornell University attended the international seminar organized by the Banaras Hindu University. Discussions with Sopher helped Singh deepen his inquiry from description of settlements to an appreciation of the processes underlying these patterns. Perceiving every theme of research from both an unconventional and a conventional point of view lent Singh's publications an outstanding depth of intellectual vigour. Not only was he a talented researcher but also a versatile teacher. Taking on his class with humour, digressive anecdotes and a wide canvas of theoretical discussions Singh's students never suffered a moment of boredom.

On the pillars of research and teaching Singh earned a reputation for his Department of Geography. It was during his term as chairperson from 1978 to 1993 that the department captured the Special Assistance Programme of the University Grants Commission. The fact that after his retirement the department lost this prestigious program, never to regain it is proof of Singh's abilities and personality. Tall and imposing Singh's gait and voice could mistake him for a military colonel but beneath the veneer of a forthright and at times candid tone, lurks a person interested in the finer understanding of social processes and problems. It is no wonder that he calls himself a social and cultural geographer with additional interest in economic geography.

Singh enjoyed his post retirement years in active teaching at the College of Social Sciences in Addis Ababa, the capital of Ethiopia, and completes his book *Dynamics of Economic Geography and World Economy at the Cross Roads* – he attributes his success to the people he met in this world of chance – if the postman had not lost the admission ticket, if educated paternal uncles had been unavailable, if he had not met Brush and Sopher, if he did not publish in the Annals, if he had not read sociology, if his department had not been recognized by the University Grants Commission, if he had not got the assignment to teach in Ethiopia – life for Singh would have been different. "If" is the middle word of life! (1932–2013)

Shah Manzoor Alam

Manzoor Alam is buried neck deep in the study of the Dead Sea Scrolls. Scrutinizing the English translations of these Hebrew and Aramaic texts, Alam's aim is to decode the principles which unite religions that are contained in these works. It was a write up in *Edit*, a newsletter of the University of Edinburgh, in 1997 that led Alam towards the historical richness of these traditional texts. Identifying and creating a unique niche of work is not new to Alam. When Alam joined as a lecturer at Osmania University in 1952, the Department of Geography was languishing in obscurity. It was a department tucked way out in the south and stood isolated from the heartland of geography which held sway in northern India. Intellectually the university was impoverished as 60 to 70 per cent of its faculty had fled to Pakistan in the aftermath of the Partition of India. It is here that Alam built a department which the University Grants Commission recognized as a Centre of Excellence way back in 1978. That is not all; in 1982 Alam founded the Centre for Area Studies. Its objective to bring together the countries of Indian Ocean to a common centre for research was unprecedented not just in Andhra Pradesh and the University of Osmania but in India as a whole. Invited to a vice chancellorship at the University of Kashmir, Srinagar in 1984 was another area where Alam displayed talent. In the exceptionally volatile political challenges which embroil Kashmir Valley, Alam introduced half a dozen new courses of study and revitalized many which were virtually defunct. His vision was not just to remain in the chair, but to use his position to foster close ties between Kashmir and the rest of India. A talent promotion scheme, where more than 20 teachers from the university were funded each year for fieldwork in any part of India in the three months of winter holidays, is still remembered for its usefulness and popularity. So laudable were his works with the World Bank from 1982 to 1997, that late he remained a consultant on their files.

How did Alam bring his obscure department to a front rank position among geography departments in India, create a Centre for Area Studies, join the files of the prestigious World Bank and attract an invitation to become vice chancellor of a university in the most politically sensitive state of India?

Alam has no intellectual mentor in India, he has a doctorate from the University of Edinburgh. Born on the first day of the year in 1928 in the small town of Ghazipur, 73 km north-east to the city of Banaras, Alam does not hail from an elite family of political, economic or academic repute. It was at St. Andrew's College at Gorakhpur that Alam attended high school. Academics was not one of Alam's first career options, he wanted to be commissioned as an officer in the forces. A graduation and post-graduation from the University of Aligarh in 1948, did not add feathers to his cap. Alam did not score even a high first division to attract recognition. A short year as lecturer at an intermediate college in Jaunpur in 1949 and a term of teaching at the National Defence Academy were not launch pads of any sort. Filled with pride and dignity Alam is far from the type who networked or begged favours. What, then, was the chemistry to Alam's success? Innovative research in geography is what catapulted Alam into the stage spotlight.

"Can the models of the evolution of the American city propounded by Hoyt (1939) and Harris (1945) explain the characteristics of the Asian city?" was the interesting question Alam addressed in his doctorate. His results are truly penetrating, very revealing, and [an] original contribution to urban geography is what W. Watson writes in the preface of Alam's published thesis: *Hyderabad – Secunderabad – A Study in Urban Geography*. Availing himself of the British Council travel grant and the University of Edinburgh scholarship it was under the tutelage of Watson, chairperson of the Department of Geography, Edinburgh that Alam completed his doctorate in 1962. Alam's use of a mixture of theory, common sense, and inferences drawn from data challenged the American models and led him to the conclusion that all cities are too much the children of their cultures to be alike, each speaks in a unique way of the history and geography of the region it expresses. Studies on urban geography were not new to the academic landscape of Indian geography, but Alam's work was a departure from the hackneyed urban studies which were prevalent in the country. His research so stirred the academic community that Alam could not be ignored. Invitations for visiting professorships poured in from as far as the Massey University in Palmerston, New Zealand and the Macquarie University in Sydney, Australia where he spent a year each in 1970 and 1974. The *Planning Atlas of Andhra Pradesh* (1976), the first of its kind in India, was Alam's next turf for success. Bringing together three diverse departments: the State Planning Board of Andhra Pradesh, Survey of India and the Department of Geography, at Osmania University, the thrust of the atlas held a strong message – administration can understand planning better through maps. The success of the atlas invigorated Alam towards another unusual creation – the Indian National Cartographic Association in 1978. It was Alam's brainchild. When Joy Dunkerley, the economist with the senate of America was in search of an anchor for her project on fuel wood in urban markets, Alam was an obvious choice. This paved the way to a series of projects on urban energy with the World Bank through 1994 to 1997. Writing five books, editing another ten and publishing 50 research articles, it is in the harvest of innovative works in geography that Alam reaped his bounty of success.

Creative people are sensitive species who tend to retreat rather than adapt when the environment gets hostile. Even though nurtured with personal care when conferences turned more political than professional, Alam preferred to limit his days of attendance. When political pressures interfered with his vision of becoming vice chancellor, Alam forwarded his resignation.

How does Alam peg himself? Reticent and humble he attributes his works to the will of God. Little wonder that Alam has found another unique niche – to bring alive the secrets of the Dead Sea Scrolls. (1928–2021)

Rajagopala Vaidyanadhan

Vaidyanadhan has to his credit the interpretation of over 1,000 aerial photographs and numerous satellite imageries. Yet focusing rapt attention on interpreting an aerial photograph through the stereoscope is not Vaidyanadhan's main interest. He has documented his research in 70 articles, published in a variety of geographical and geological journals. Today he is busy completing his book on the *Geology of India* (2010). Yet, hankering after publications is not Vaidyanadhan's passion. He has led field visits to places as far reaching as Pahalgam in the north to Cape Comorin in the south, Bhuj in the west to Tezpur in the east. He has also conducted geomorphic mapping camps to train officers from the Geological Survey of India. Fieldwork alone is not a burning obsession with Vaidyanadhan. With patience and care he has edited the *Journal of the Geological Society of India* from 1992 to 1995. Editorial works too do not hold fascination for Vaidyanadhan. For many years he served as member of the governing body of the Wadia Institute of Himalayan Geology and has been an expert on committees for the Department of Science and Technology, University Grants Commission and the Centre for Scientific and Industrial Research. These, he says, are academic chores which are best handled in small dosages. As a rule, Vaidyanadhan has stayed away from a variety of ships – principalship, wardenship or vice chancellorship. Despite a first-class first in school and at both undergraduate and postgraduate levels, Vaidyanadhan did not opt for a career in the Indian Administrative Services. Making decisions on the grounds of administrative logic was not his strong suit. What, then, has been the energizing force which propelled Vaidyanadhan through an uninterrupted innings of 38 long years at the University of Andhra Pradesh? "Students,"' is Vaidyanadhan's prompt reply.

The answer discloses the secret of his being a teacher of established repute. Every year the university syndicate of Andhra Pradesh selects one out of 70 odd professors for the Best Teacher Award. In 1983 in a grand ceremony held at Hyderabad, in the august company of the state chief minister, vice chancellor and other dignitaries, Vaidyanadhan's name was announced for this prestigious award. Vaidyanadhan was busy undertaking fieldwork with a group of students. The award was given in absentia yet Vaidyanadhan's roster for the day had in a way kept the spirit behind the award alive. When the academic committee of the university proposed the inauguration of a department of geography, Vaidyanadhan was their first choice. Serving in the department of geology, it was not just the fact that Vaidaynadhan's doctorate was on Geomorphology of the Western Cuddapah Basin, but more his reputation as a teacher that weighed in his favour as an obvious candidate. A devoted teacher, as soon as Vaidyanadhan switched from geology and joined the Department of Geography in 1973 his first task was to attend to the needs of students.

Today if the geomorphology lab in Waltair University's department of geography is proud to possess 200 pairs of stereoscopes, 250 aerial photographs and a dozen imageries, a considerable share of the credit for this goes to Vaidyanadhan. Inviting experts from the fields of oceanography, climatology and hydrology to the Department of Geography, was Vaidyanadhan's way to inculcate an interdisciplinary culture of learning. More important than infrastructure was the judicious blend of geology and geomorphology which Vaidyanadhan grafted into the department. In spite of being a geologist for 20 long years, Vaidyanadhan never overlooked the fact that he was to service the needs of his students. Sensitive to the reality that most geography students belong to the humanities and not the science stream, Vaidyanadhan focused his attention on the study of deltas, coastal and erosional surfaces where knowledge of geology is needed but only at an elementary level. Clarity, commitment and an ability to communicate is what made Vaidyanadhan a teacher with class. Focusing on processes rather than patterns, principles rather than description, a doctorate student of his summed up with him one knows where one is. Locating oneself is, after all, a critical step in moving ahead. Combining the skills of the spoken word with visual aids and verifying them with reality in the field soaked Vaidyanadhan's lectures in power. Most versatile were his ways to teach the interpretation of aerial photographs and imageries.

Vaidyanadhan had availed of the Smith Mundt Fulbright grant to study photogrammetry and photo interpretation at Ohio State University, Columbus, USA in 1959–1960, but he had a gifted knack of visualizing flat features in a three-dimensional perspective. To Vaidyanadhan there was no flat surface, all land had a concavity or convexity, and at times both. He went out of his way to train students to develop this eye. Procuring aerial photographs in India is ridden with problems of secrecy and bureaucratic red tape. To overcome this lacuna, Vaidyanadhan took on ten research projects under the banner of different sponsoring agencies such as the Oil and Natural Gas Corporation, Ministry of Defence and Department of Science and Technology, so that photographs were readily accessible. He personally organized visits for his students to the Photo Interpretation Centre, Dehra Dun; the National Remote Sensing Agency, Hyderabad; and Geological Survey of India, Calcutta. Students were the end users of his research. While Vaidyanadhan published in both geological and geographical journals, he was careful to craft his papers according to the audience. The papers on structures, fractures and lineaments sought an outlet in geological journals while those on landforms, land use and applied geomorphology were directed to geographical journals. Nearly 50 per cent of his papers were written with students. Training and encouraging research scholars to publish, one of his hallmarks is that Vaidyanadhan was generous to give credit to them. A check of his publications shows that the research scholar is the first author and Vaidyanadhan the second. This humility earned him deep reverence among his students.

To Vaidyanadhan, venturing far from his department was a colossal waste of time. A majority of his research papers are also rooted within Andhra Pradesh. He explains that a juxtaposition of mountain, coast, and delta in close vicinity is ample opportunity to test processes and patterns and relationships of the physical world. Though he was born on 21 December 1931, in the Kallakurichi village of the South Arcot district in Tamil Nadu, most of his education and all his professional life has been spent in Andhra Pradesh.

A distinguishing trait of a teacher is to continue learning and prepare for the next class. Residing, after professional retirement in 1991 in a small district town called Cuddalore 22 km to the south of Pondicherry, Vaidyanadhan, a connoisseur of Karnatak music is fully engaged in learning to read and recite the Vedic shlokas and scriptures. With unflappable detachment so typical of many South Indian Brahmins, he states: "I am 70 years of age, the average life expectancy in India is around 62, I am living on a bonus and therefore preparing for the 'next world'." The sequential progression: facts, analysis and conclusion in this sentence alone epitomizes the teacher in him. All who join the academic profession teach. But there are few teachers. Vaidyanadhan is one of them. (Born 1931–)

Satyesh Chandra Chakraborty

In a problem ridden world those who provide solutions are at a premium. Satyesh Chandra Chakraborty provides solutions to problems. Among the types of issues Satyesh has tackled are water disputes of Bihar, Orissa and West Bengal, policy imperatives in Calcutta's Master Plan, designing expressways and prioritizing funds for different infrastructure. The spectrum of audiences who sought advice at his doorstep range from the Calcutta Metropolitan Authority, Calcutta Municipal Corporation, Reserve Bank of India, Pollution Control Board, authorities at the tea estates and mining groups, and solid waste managers. His address list of 57 clients draws private and government concern from places far and wide. Though the state of West Bengal leads with a 50 per cent share, ten other states including Maharashtra, Karnataka, Kerala, Haryana, Rajasthan and Madhya Pradesh can be counted straight away. Outside the shores of India Satyesh has attracted seekers from the Middle East, Africa and the former Soviet Union. These buyers are not an easily satisfied lot. Many demand more than a pound of flesh. Take Georgia, a former Soviet Union Republic covered by the rugged Caucasus Mountains. Its harsh climatic conditions would have daunted many, but not Satyesh. Reaching 70, an age when most are ossified, Satyesh learnt the state-of-the-art technology such as geographical information systems to raise queries before reaching an appropriate management decision, be it to lay a road or design a city health plan. A hefty pay cheque which customers use as bait is not the only temptation; the challenge to solve problems galvanizes Satyesh. In modern parlance Satyesh could professionally be called a consultant. In a traditional context those like him would be seen to possess a canvas of mind vast enough to comprehend problems and elicit solutions.

Satyesh possesses a doctorate in the discipline of geography. Does he provide geographical solutions? The answer is yes and no. Yes, because he is a geographer and no, because Satyesh is not only a geographer. He is a polymath. His interests cut across a variety of disciplines – sociology, economics, political science, social psychology, history, calculus and astrology.

Satyesh has a keen mind to learn, exceptional powers to memorize and an ability to integrate a range of phenomena when drawing upon an interpretation. His special ken is decoding processes behind patterns. Little wonder that the Department of Geography at the Presidency College could not hold him permanently or the University of Burdwan could not retain him for more than four years. An invitation to join the Department of Geography at the Delhi School of Economics was also turned down. Satyesh chose to be a professor and chairperson at the Indian Institute of Management, Calcutta where he served from 1974 to 1991. He did not take an extension of service and refused an offer to join the University of Calgary; on the contrary he engaged himself full time in private consultancy. That 70 per cent of his total advisory works fall in the phase of his private consultancy and he is on the rolls of many firms are proof of his credibility. Satyesh's ability to solve ticklish problems is backed by an almost hypnotic power to convince. His 91 research papers cover a range of themes that defy classification. Though scholarly on many levels, he is at his best with the spoken word. He has a passion to be heard. His 30 public lectures on themes like marine policy, strengths and weaknesses of cooperative workers, and the informal sector in India to Calcutta's urban space have seen more of an international podium. Three-quarters of his public lectures were aired at universities and research centres such as East-West Center in Honolulu, the Free University of Berlin and Dresden University in Germany, the University of Chicago and the London School of Economics., A tenure Scandinavian Universities has also been on his cards. Few who have heard Satyesh can escape the spell of his oratory charm. Borrowing a clipped British accent, through a dense haze of Havana smoke, brick by brick Satyesh builds his concepts and lays across his arguments. His words trap one and all. He is not a laden tree that stoops. His girth is strong and he bears his weight in apparent conceit. Yet he is not toffee nosed and bends backward to learn. He not only secured a first-class first in geography from Presidency College but also bagged the title of Eshan Scholar for the highest rank in the faculty of arts. In his post-graduation he won gold medals all the way; for top rank in geography, for the highest rank in the faculty of arts and the highest rank in practicals among the science faculty. Although the University of Calcutta gave Satyesh honours, he reminiscences that his school days were the grand period of learning in his life.

Comilla, a town to the south of Dacca in Bangladesh, is where Satyesh was born on 3 November 1931. The transferable job of his father, who was an advocate in the British judiciary, meant that Satyesh's childhood was punctuated by a dozen schools in different places. Far from feeling uprooted it built in Satyesh a resilience to cope with new situations. From the JBD ink tablets to the digitizing tables, for Satyesh technological, methodological or conceptual shifts have never been a barrier.

"Beyond the mountains there is a plateau, from where three mighty rivers originate, Lhasa is its capital – where is Tibet?" Satyesh recollects this as his first geographical question, but he affirms that the interest in natural sciences unlocked his eagerness to study geography. There were years of self-learning at Presidency College where the bazaar of book shacks on the perimeter wall was frequented more eagerly than lectures in the classroom. A heightened intellectual activity came to Satyesh at the London School of Economics. A joint supervision under Oligivetee and Rawson for a doctorate was another strong impact.

Geography, Satyesh concludes, has tremendous scope for being marketed outside academia, only geographers have to root their discipline in sense. That he quips is not so common among the community of Indian geographers. Since Satyesh has identified this problem – perhaps he should be the one to find a solution. He took his time to plough back into the discipline from which he has earned so much. After all, his talent to memorize and interrelate a range of disciplines, experiences and learning is what makes Satyesh a consultant par excellence. (1931–2011)

Vidhiparti Lova Surya Prakasa Rao

In a private room of the Medwin Hospital at Hyderabad was raised a unique appeal, "geographers in India must research if the downhill path of the discipline is to be arrested." The words carry weight. For they emanated from the heartfelt anguish of one of India's grandmasters of geography – Prakasa Rao. On 22 December 1992, grief filled the room as the frail gasping body ceased breathing.

The talent for conceptual and methodological clarity, an analytical ability backed by a plethora of creative ideas is what distinguishes a master from the rest. Prakasa Rao mastered the craft of the discipline of geography. He was among the first Indian geographers to use factor analysis in 1954, almost a decade before Berry popularized this technique in the USA. He questioned the critical role of small and medium towns in regional development two decades prior to such a realization on the part of the United Nations Development Agencies. Prakasa Rao evolved a composite index for groups of regionally differentiated characteristics and chose themes like central places, the role of growth poles, and growth focuses to propel an in-depth inquiry into regional development as early as the 1960s.. This is not all. The locational analysis of inter-urban patterns, the changing characteristics of the rural–urban continuum, and the reorganization of administrative units, all are ideas which can be traced to the prolific mind of Prakasa Rao. He did not simply broadcast ideas but had the intellectual grist to test and validate his hypothesis. He disliked simplistic descriptive conjectures and he fuelled his analysis on rigorous quantitative techniques before formulating a conclusion. It is on the shoulders of applied statistics that Prakasa Rao identified ways of measuring centrality, investigating disparity and studying spatial structures. An early brush with the data made available by the Census of India convinced him of the limitations of secondary data and it is on the girders of primary data that Prakasa Rao rested his analysis. His relish for fieldwork is exemplified by the village level data for the study of the Muzaffarnagar district, the socio-economic questionnaire for the survey of Bangalore, and the 1,075 household interview schedules generated at Vijayawada. Drawing experience from the National Sample Survey, Prakasa Rao aimed at designing a spatial sampling frame to accomplish a regional trend analysis. Selecting areas of study, which range from the micro- to meso-scale allowed Prakasa Rao to draw meaningful conclusions at the macro-level. Furthering an original idea, applying a novel technique for analysis, identifying new methods of data collection were not the end of his research. A hallmark of a master is one who likes to present and arrange his results in a meaningful and near perfect manner. The map was Prakasa Rao's favourite tool for display and his adept skill at cartographic techniques gave him exceptional power to access designs and styles. The *Atlas of the Urban Landscape of Vijayawada*, as an example, exhibits a stylish combination of lines, shades and symbols whose insets and superimposed graphs lend an aesthetic quality to its 24 different plates. All masters sharpen their tools before they chisel their ideas. Prakasa Rao was no exception.

Born in 1917 in the town of Visakhapatnam on the east coast of India, Prakasa Rao was the eldest child in the family. Completing his education at the CBM Government School, Prakasa Rao enrolled for a Bachelor of Commerce at the AVN College at Visakhapatnam. A course on economic geography in the curriculum seems to have turned Prakasa Rao away from commerce and towards geography. Completing his post-graduation in 1942, he plunged headlong into a doctorate from the University of Calcutta in 1950.

It was his apprenticeship at work rather than formal education which became a playground for testing ideas and techniques. Prakasa Rao's fervour for innovative research made him pen articles on town planning, the location of industries and the scope and limitations of land use surveys as early as the 1950s. The assignment as editor of one of India's oldest journals, the *Indian Geographical Journal*, gave Prakasa Rao an additional advantage of access to the literary world. But deep at heart Prakasa Rao longed to expand his creative impulse particularly in the area of urban and regional planning. His intention found manifestation. Prakasa Rao was selected to join the regional planning survey of the state of Mysore, under the auspices of the planning commission. Working intimately with Learmonth and Mahalonobis, Prakasa Rao's keen intellect outshone his peers and when the Regional Planning Unit was set up at the Indian Statistical Unit in 1956 he was appointed as its first director.

In the search to master a study many masters become footloose. Prakasa Rao's spurts of dislocation from Calcutta to Madras back to Calcutta, then to Hyderabad to Delhi, and at last to Bangalore highlights a soul in search of a perfect environment for research and study. Prakasa Rao could not retain himself in any institute for more than seven years. His stay at the Indian Statistical Institute lasted for five years, he was a professor in the Department of Geography, Osmania University for an interlude of five years, and the Department of Geography, Delhi School of Economics for seven years. It was at the Institute for Social and Economic Change at Bangalore that Prakasa Rao where spent another seven years, with the last four years of his professional life being spent at the Indian Institute of Management. A brief teaching assignment at the University of Syracuse, US, was also explored as a place of possible stay.

In the seven year periodicity of movement, Prakasa Rao's interactions with statisticians, economists, sociologists and those from the management field fertilized his thinking and whet the tools of his inquiry. Though brief, each station at which he stopped reaped abundant benefits from the research led under Prakasa Rao's sharp eye. *A Survey of Towns in Mysore* at the Indian Institute of Statistics, the *Study of Vijayawada and the Regional Structure of Muzaffarnagar District* at the Delhi School of Economics, and the *Study of Bangalore* at the Institute for Social and Economic Change in the city to date remain monumental pieces of work. Masters give back more than they take.

A pious and humble soul, Prakasa Rao did not want followers but original researchers. Prakasa Rao, a scholar who played the harmonium and filled his study with notes from the violin of Yehudi Menuhin, is no more. To heed the words of wisdom that Prakasa Rao whispered on his deathbed would be a tribute to this grandmaster of geography from India. (1917–1992)

Laxmi Niwas Ram

Ram lived in an exile of circumstances which thwarted his talents to blossom to the fullest. Educated at the elite St. Xavier in Patna, the capital city of Bihar, securing an unbroken record of first divisions from secondary, intermediate, graduation to post-graduation levels are test grounds of Ram's intellectual faculties. Awarded the Patna University Merit Scholarship in 1956–1958 and the Patna University gold medal in 1958 cements this fact further. Being selected as a lecturer at the young age of 23 at the Department of Geography, Patna University in 1960, securing a professorship in 1984, and his appointment as an acting vice chancellor of Patna University in 2001 demonstrate the many heights to which he rose in his academic career. An impeccable flair for language, keen analytical ability and carefully cultivated articulation are his other assets. A personal collection of detective and science fiction and classics of English literature remind us that Ram's avid interest in books extended beyond geographical texts. He enjoyed carving sentences and weeding verbosity – this can be seen from the 24 years he spent in an editor's post, initially for *Indian Geographical Studies* from 1973 to 1990, followed by the *Geographical Perspective* from 1990 to 1997, both journals of the Department of Geography, Patna University. His consultancies in the capacity of member of the Master Plan for Patna and the Sone Command Area Development Agency are proof of his competence in the applied role of the discipline. Ram's efficiency to inspire and weld the spirit of teamwork is amply displayed in the number of chores that he took on. Take, for instance, the Indian Geography Congress of the National Association of Geographers, India, which in 1991 was under the umbrella of his presidentship. Ram recollects the spirit of cooperation he could instil in his fellow colleagues as chairperson of the department and president of the National Association of Geographers, India to collectively bring forth a publication on *A Systematic Geography of Bihar*. Serving simultaneous posts as Dean of Faculty of Social Sciences and Acting Principal Patna College, he dared to take on challenges. Ram reinitiated the mid-term examinations which, as a result of student protests had been abandoned for over 10 years in Patna College. Though holding a doctorate in Urban Geography, Ram holds the firm opinion that, "geography cannot be segmented into specialisations – embracing an understanding of all its themes is the strength of the discipline." Teaching a wide menu of courses in geography from cartography to physical, and political to urban geography affirms that he put this belief into practice. The 30 doctorates supervised by him also spill into themes as far ranging as settlement, political, industrial, agricultural, resource and urban geography. He also does not shy away from experimenting with innovative techniques and has been on the advisory committee of a study in Bihar where geographic information systems were an important tool for analysis. Cosmopolitan in taste, his hobbies range from tennis and photography to electronics. Residing in a sprawling *haveli* in the centre of Patna, on whose perimeter he owns two dozen shops, speaks of opulence uncommon for most academicians in India.

Despite all the medals, ranks, skills and roles, Ram's intellectual merit outranks his performance. Most of his works are confined within the territorial limits of Bihar. His entire education, all 37 of his teaching years, 90 per cent of his supervised doctorates, all his four research projects, five out of the nine articles he has published, both of his edited books, are on Bihar. A curious question remains as to why Ram has no legendary publication to his credit and why he has not been able to fulfil his very own professional desire to produce a geographical treatise on Bihar.

Born in 1937 to a family of zamindars, the early death of his father left Ram with a wealthy estate but also chained him to a retinue of litigations to be handled in the courts.. Though Ram's heart lay in the intellectual pursuits of the discipline of geography, yet the pressing demands of the court and the tax rooms stole his peace of mind, Yet Ram remains an optimist and says that the book *Geography of New Bihar and Jharkhand* will soon see the light of day. Circumstances make a person but a person can also create his own circumstances. (1937 –)

Padmanabha Doodhbhate Mahadev

Carrying a near permanent demure tone and an unflappable temperament, it is difficult to comprehend the element of dissatisfaction that Mahadev harbours on the academic front. To all apparent observation Mahadev's professional career has numerous ingredients for satisfaction. He has successfully climbed the professional rungs from lecturer to professor in the Department of Geography, University of Mysore. To his credit is the supervision of ten doctorates, and 14 research articles in eight different geographical journals, both in India and abroad. He has chapters in 17 different books and has authored or co-edited another eight books. As principal investigator he has undertaken four research projects, half of which have had foreign funding. This is not all. Mahadev has attended over 15 international conferences and has been invited on research or teaching assignments to several universities outside India. The Western Washington University and the University of Pittsburgh in the US, the Talence Campus at the University of Bordeaux in France, the University of Kyoto and University of Hiroshima in Japan, and the University of Groningen in the Netherlands are all listed in his curriculum vitae. Besides the National Association of Geographers, India, Mahadev has been president of two of India's other important leading geographical associations, the Indian Council of Geographers in 1982, and the Indian National Cartographic Association in 1993. Over and above, Mahadev's self-effacing and humble countenance has won him friends everywhere. Despite the mileage on many fronts, Mahadev's career trajectory is punctuated by signals of dissatisfaction.

Mahadev is among the very few geographers in India who has a double doctorate in the discipline. In 1969 the University of Mysore awarded him a doctorate, while four years later he worked towards another degree from the University of Pittsburgh. Not only are both the doctorates in the discipline of geography, but both are in urban geography and both are on the city of Mysore. A single doctorate did not suffice and Mahadev affirms that the second from Pittsburgh added a little more to his wealth of exposure and experience.

A similar unease is noticeable in the professional shifts undertaken by Mahadev. After serving for a year as lecturer at the Alagappa College in Andhra Pradesh, Mahadev taught for five years at a college in Dharwad. He joined the Department of Geography, University of Mysore in 1964. After a decade and a half of teaching and chairing the department, he moved to the Institute of Development Studies at Mysore as Professor of Urban and Regional Planning and stayed on from 1979 to 1985. A brief stint of six years after which he reverted back to this parent department. The research themes he pursued also carry a search for a subject of interest. A knock down of his 14 research articles show that half are on urban geography while the remaining relate to agriculture, settlement, industrial, cultural and environmental geography. His research projects too revolve around subjects which are poles apart – from the impact of irrigation on settlement, to rural health services, to Japanese research on geography in India.

The years spent at the Institute of Development Studies gave Mahadev a spring pad for travel, contacts and to widen the vistas of his learning. Flush with grants from the Ford Foundation, the institute made it possible for Mahadev to organize numerous workshops on, for example, locational analysis and research methodology, and to attend the International Geographical Union held in Vienna and Paris in 1984. Visits to the Ben-Gurion University in Israel, the University of Groningen in the Netherlands, and the University of Hiroshima in Japan also fall in the time period when Mahadev was professor at this institute. Despite all such benefits, Mahadev took a U-turn to go back to the Department of Geography at University of Mysore.

Mahadev recalls that the choice of selecting geography as a discipline for study also had no serious base. Born on 16 August 1933 in Coimbatore, his father's transfer as an employee of the government gave his schooling an itinerant character. Completing his studies at the Board High School in Chittoor, then in the former Madras Province, Mahadev joined the Presidency College there to pursue his graduation and post-graduation. Throughout school or college, no single subject of study caught his total fascination. If today he was given a chance to choose a discipline for study, he says it would be anthropology. In fact he adds he would prefer to be in administration rather than in teaching.

What is remarkable is that despite a bed of internal uncertainties and strife Mahadev carried out his professional duties with pleasant ease. He never turned cynical and never cast a caustic comment. A recent mild paralytic stroke clouds his health. In spite of his agony, Mahadev maintains a compassionate composure. He exemplifies a noble learning of how a profession is to be pursued despite dissatisfying odds. (1933–2005)

Shanti Lal Kayastha

Daily in the wee small hours, a click-clack can be heard from the Akhnoor Hut in Varanasi. It makes two announcements – first the time on the clock is 04.00 AM and second, Kayastha has set to work on his Olivetti machine. Perched amidst his personal library – a treasure room of around 5,000 books and reports, a catalogue of photographs, slides and maps, it is clear that Kayastha's typewriter will never run out of facts, data and references. The map of his room is etched meticulously in Kayastha's memory and at the snap of a thought or a behest of a request he can pull out a reference or summon a report. Half a dozen cameras dangle from a nail on the wall, along with knapsacks and weather braving jackets. They tell us Kayastha is not an armchair geographer. Disciplined and painstaking work shouldered on extensive fieldwork are steps on which Kayastha has built his scholastic reputation. A tireless labour for geography has brought to Kayastha's doorstep heavy duty work, eager students and scholars.

For nearly three decades Kayastha was the honorary editor of the *National Geographic Journal of India*. Today if this journal has a reputation, it owes a debt to the many hours of checks and rechecks sifted singularly by Kayastha. Many among his 20 doctorate students preserve their drafts or manuscripts as pieces of learning from their supervisor. Kayastha's five books, 70 published articles and over 40 papers presented at various symposiums and conferences carry the stamp of his persistent diligence. Rewards came to him in ways both bold and subtle. The National Geographical Society of India conferred on Kayastha the prestigious Sardar Vallabh Bhai Patel Medal in 1971; while a *festschrift* titled, "Geography and Environment: Issues and Challenges," is a befitting tribute by students to their revered teacher. Little deterred to seek worldly awards, Kayastha maintains a steadfast commitment to any work he undertakes.

From 1956 to 2001 Kayastha has a virtually uninterrupted record of attending the Indian Science Congress, quite an achievement considering that he never arrived at the meet without a written paper. His work earned respect and he was elected as president of the Geology and Geography Section in 1988. Among the geographers in India he is one of only four geographers who have acquired this post. The International Geographical Union is another of Kayastha's favourable meets. Off the cuff, he can remember a string of attendance ranging from Delhi 1968, Moscow 1979, Karachi 1980, Tokyo 1980, Kathmandu 1983 and Sydney 1988. His presence earned him full membership of the International Geographical Commission on Environmental Problems and corresponding membership of the Working Group on Environmental Perception and the Commission on Population and Environment. Conferences and congresses are a mere speck in the life of a true academician. What is more endearing and enduring is the area of research. It is in the focus of research interest that Kayastha's allegiance is best at display.

The region encapsulated within the towering Dhauladhar range of the Himalayas and the Beas, one of the five major tributaries of the Indus River, became Kayastha's unstinted preoccupation for over three decades. Nearly 40 papers and one of his books completely orbit around the Beas Basin within the state of Himachal Pradesh. The attraction for this Himalayan state holds his attention to date. Every summer Kayastha can be seen walking, observing and taking pictures in Kangra, Dharamshala and Dalhousie. More than half his album of photographs captures the Beas River as it flows through snow-capped mountains. Kayastha's bond with the Himalayas run deep and long.

Born in the small town of Kangra on 24 March 1924, Kayastha spent his childhood skirting dales and vales. Walking to the Dayanand Anglo Vedic High School in Kangra versed him with the ways in a mountain. Perhaps it is Kayastha's idyllic attachment to the land and landscape that prompted the principal of the Government College at Lahore to advise this undergraduate student to elect geography as his major in higher education. While pursuing a post-graduation at Aligarh, his association with late Muzaffar Ali ignited in Kayastha a desire to undertake a doctorate from the UK. A letter of registration with Dudley Stamp at the London School of Economics in Kayastha's personal file recounts a tale of dreams and sacrifice. Kayastha opted to share the burden of rearing a family of nine brothers and sisters with his father, rather than take off on a personal rendezvous. A twist of destiny which perhaps played a role to take him back to the lap of his first love – the Himachal Himalayas.

When Kayastha joined Banaras Hindu University as a lecturer in 1948 and set to work on his doctorate the mountains were an obvious choice for study. He spent seven devout years conducting fieldwork and assembling data in the region. Many crooks and crannies were scrupulously scanned by Kayastha as he spread his fieldwork over 25 villages selected across different habitats of the Beas Basin. Published in 1964, Kayastha's thesis was hailed as pathbreaking on many fronts.

Most significant was the unit of study – a basin. In the 1940s the concept of a river basin as a functionally unified system of study was little known to geographers in India. It was a chance meeting with A.T. Morgan, President of the Tennessee Valley Authority, that stirred Kayastha's imagination to choose a river basin as an area of study. Following perhaps Geddes' principle of folk, place, and work, Kayastha's research titled, "The Himalayan Beas Basin: A Study in Habitat, Economy and Society," won immediate attention.

A prolonged and consistent eyewitness of one region sensitized Kayastha to the changing environmental scenario and the menace of degradation and pollution in the Himalayas. The processes and patterns of deterioration in the mountains reverted Kayastha's attention to the widespread environmental issues across India and the earth as a whole. Kayastha is unwilling to take the credit of his role in popularizing environmental studies among geographers in India. He humbly acknowledges Gilbert White, emeritus professor at the University of Colorado as his intellectual mentor. Kayastha personally met White only twice; once in 1968 at the International Geographical Union Congress in Delhi, followed by a second meeting at the congress in Tokyo in 1980, yet a stack of letters with Kayastha is proof of the exchange of news and views between the two scholars. The introduction of a course on environmental geography in the curriculum of the Department of Geography at Banaras Hindu University and a series of public lectures on the interface between environment and development, environment and population and environmental policies are planks on which Kayastha evolved his ideas and identity. Despite his faltering eyesight Kayastha has recently finalized a paper called *Environment and Development: Crises and Conflicts*, for publication. A comparative account of environmental problems in countries he has visited is his next agenda.

From the crack of dawn to dusk, day in and day out, it is the mould cast by a discipline of hard work which drives Kayastha at the age of 77 to accomplish a well-defined journey ahead. (1924–2018)

Parmeshwar Dayal

Among the handful of doyens in Indian geography – Dayal stands apart as a legend in his own time. Conservative in manners, aloof in style and thoughtfully weighed in speech and action – Dayal commands outright respect. His prestige is based more on merit than position. Chairperson at the Department of Geography, Patna University for three decades, few among his colleagues can recall an incident or decision where Dayal faltered or stooped. Serving the cause of the discipline, Dayal's meticulous punctuality of being in office from ten to five set a culture of work. Although not the founder of the department, as its first chairperson, he irrigated its various needs with rare efficacy. That many of the alumni of this department today occupy esteemed positions in India and abroad is a result of the high standards of education they gathered at their alma mater. If the Department of Geography at Patna today refers to a glorious past, the illuminating light goes to none other than Dayal.

Creating a disciplined atmosphere for teaching and research, Dayal fostered an exemplary solidarity within the department and between faculty and students. Assured of a cooperative working spirit, Dayal netted credos in numerous ways. Research projects from the Ministry of Defence, Planning Commission and the Indian Council of Social Science Research being one among many. These paved the way for the department to be recognized as the Centre for Development Studies under the auspices of the University Grants Commission. A personal persuasion to set up a number of departments of geography in different parts of the state of Bihar earned Dayal additional respect. When Bihar formed its first State Education Board in 1975, Dayal was its first chairperson. Among a string of his predecessors and contemporaries Dayal stands out as the first geographer to be honoured with the position of vice chancellor in 1975. It is different that he chose to resign within a span of two years rather than buckle under political pressure. The hallmark of an exalted person is discerned when he testifies that prestige is more important than post.

Upholding integrity, a shower of positions aggregated towards Dayal. The Simmon Fellowship took him to the University of Manchester as a visiting lecturer in 1956–1957. He is one among four geographers in India who have been elected as the President of Geology and Geography Section of the Indian Science Congress. His association with Patna University was diverse in expression: eight years as president of the Patna University Teachers Association, ten years as director at the Indian Administrative Services Coaching Centre, two terms as Principal of Patna College, and five years as editor of the Transactions of the Indian Council of Geographers.

Dayal's presence lingered at places where he worked. This is evident from the experience at the University of Sri Lanka in Colombo, where Dayal served as lecturer for a short stretch of 18 months during 1948–1949. The vice chancellor, Iver Jenning, sent repeated requests for Dayal to return. He did not oblige. A visit to the University of Canberra, Australia opened another set of similar opportunities. Again Dayal declined – Patna benefited. Patriotism too is a hallmark of the noble.

Canons to the scholastic regard which Dayal drew is evident in *Encyclopaedia Britannica* (1974, pp. 984–988) which carries a write up on the Geography of Bihar, authored by him. "Dayal's analysis of the correlation of soil, climate, and topography with harvests in Bihar is extremely interesting, unfortunately too detailed to summarise." A note in the book, *India and Pakistan: A General and Regional Geography* by Spate and Learmonth is another tribute to him. Cherishing quality over quantity, Dayal's list of publications are a modest, but meaningful number. Five out of the eight books authored by Dayal are texts. Simple titles like *Commercial Geography*, *An Outline of Practical Geography*, *Study of Modern Geography*, and one on *Geomorphology* are of direct utility to school students and for those at the undergraduate level. Most books are in Hindi and thus confirm that Dayal gave priority to the needs of the students at the cost of his own research interest. Belonging to the traditional school of writing, Dayal had a disdain for edited and co-authored works. Sixty per cent of his books and 85 per cent of the research articles are authored by him alone.

Born on 15 August 1920, in Bind village in the Nalanda District of Bihar, Dayal's ability to stay attached to his books all day long did not go unnoticed by his parents. When the day to finance Dayal's education to England arrived, the family dipped into its savings with a matter of pride. Boarding an overloaded troopship which housed 5,000 rather than its usual 1,000 passengers meant that Dayal's journey from Bombay to Liverpool was far from pleasant. But for Dayal these were trivial concerns. The war-ridden Britain where food was rationed and inflation ran amuck were also taken in stride – the London School of Economics, geography and a joint supervision under Dudley Stamp and OHK Spate were all that mattered.

An event in history had taken a turn in 1947. When Dayal left, India had been a British colony; he returned to a free India. The enthusiasm to dedicate and build the nation was on high ground. Dayal had a legendary mission: to build and establish the discipline of geography on his home turf. (1920– 2015)

Laxminarayan Sanna Bhat

To swim snugly in a swarm of one's own professional community may have been the desire but was not Bhat's destiny. Though a popular professor of geography, author of 13 books, several research papers in leading reputed international and national geographical journals, and contributing author in *The First Survey of Research on Geography in India*, a member of leading committees which evaluate and assign projects to geographers – Bhat does not typify a common academician in a university. His career is testament that even when positioned outside a department of geography while at the Indian Statistical Institute one can build one's reputation as a geographer.

Joining the Regional Planning Unit at the Indian Statistical Institute as a researcher at the age of 28, rising to the rank of professor at 45, Bhat spent 32 long years in this institute as a regional planner. Bhat belongs to the very small group of geographers in India who practise what they teach. The tenets of area and region for Bhat were not mere conceptual and methodological tools of reference for lectures in the classroom and critiques in reviews, they had to be used in the field, incorporated in projects and implemented through plans. The test of any plan, after all, lies in its implementability. Perfecting the cuisine of regional planning for Bhat has been a lifetime commitment. Almost all his research projects, 80 per cent of his 13 books, 65 per cent of his 48 research articles, four of his supervised doctorates, and the majority of attended conferences pivot around the theme of regional planning. Injecting more than just a flavour of geography into planning was an uphill task. The kitchen, one must remember, was crowded with economists, sociologists, statisticians, architect planners and regional scientists. Yet today Bhat's works on regional planning find reference among all these social scientists, and his papers have been published in journals as far ranging as *Economic Geography, Journal of Regional Science, Journal of Institute of Town Planners, Sankhya, Yojna,* apart from geographical ones such as the *NAGI Annals* and the *Bombay Geographical* magazine. Bhat has made contributions at the Planning Commission, for the Regional Development Plan for Western Ghats, for the State Planning Board Kerala, for the Town and Country Planning Organization, among other eminent organizations in India.

With geography as a subsidiary subject for his Bachelor of Science degree, and a focus on geomorphology and land use at the level of Master of Science, Bhat lacked even an elementary schooling on statistics and economics. Further, he had no formal training in regional planning His doctorate *Some Aspects of Regional Planning in India* dates to 1965, nearly seven years after he joined the Indian Statistical Institute. Serving as a lecturer at the Ruparel College of the University of Bombay in the mid-1950s, he taught the traditional subjects of the discipline which made marginal reference to planning. Despite all these limitations, Bhat was handpicked to join the Indian Statistical Institute as a researcher. How did Bhat secure a foothold within the ambit of planning in India? Many would jump to the conclusion that when the regional planning movement swept through India's development plans in the 1960s and 1970s, Bhat was on the scene. Others could add that the main advocate of regional planning and founder director of the Indian Statistical Institute, Mahalanobis was 'known' to Bhat. These are, however, remarks from those who don't know. The more piercing question is how Bhat was recognized or familiar on the scene. A simple Brahminian outlook with the adage hard work pays is Bhat's very lived philosophy.

Bhat was born on 28 August 1930 in Holangadde, a village so small that it did not have a threshold to support even a primary school. The nearest school could be found in Kumta, a port town, brought to fame by the British for its export of cotton. Along with his three brothers and a sister, Bhat walked 6 km each way to learn his ABC! Exploring different paths and observing the landscape en route became such an obsession that Bhat began to frequently write travel accounts that were published in the school magazine. A more lasting reward was his dissertation for his Master's in Science. To the students of today this may seem a surprise, but a year-long intense regional study on the Upper Panchaganga River, a tributary of the Krishna, is what won Bhat immediate attention. It was the appreciation from Spate, his external supervisor for the master's dissertation, that catapulted Bhat into the profession of planning. While laying down the blueprint of regional planning, Mahalanobis would often consult Spate. When a team for an experiment with regional planning in India was drawn, Bhat was in the list. It is his work which earned him the Leverhulme Research Fellowship for research at the University of Liverpool in 1962–1963. Hard work pays but needs replenishment over a lifetime.

Bhat integrated the spatial element into planning in different ways. Research on Mysore in the south, Karnal in the north, Ghats on the western shores of India, and queries ranging from regional disparity, regionalization and regional development have been Bhat's frequent points of interest. Visiting planning departments in France, Germany, Denmark, Sweden, and the Soviet Union have been Bhat's ways to gain comparative insight into the merits and demerits of regional planning. Working with interdisciplinary teams sharpened his understanding of the role of a geographer in the planning process. Amidst his multifaceted works, Bhat wrote books and research articles to nourish the discipline of geography. His publications are recommended in the curriculums of most departments of geography and students seek advice from him on a range of issues on regional planning. His forthcoming book *Geography and Development Process: A Spatial View* to which he is adding the final editorial touches, is a much awaited text. That he did not belong in the comfort of a single department of geography has been a blessing in disguise. The discipline of Indian geography has many a faculty confined in university departments; geographers like Bhat are indeed few. (1930–2013)

Rameshwar Prasad Misra

If an opinion poll were to be conducted to elect the most popular geographer in India today, Misra is certain to score very high. An index to his more than ubiquitous presence are citations and trophies of his achievements: a Fulbright Fellowship, the Golden Jubilee Award of Mysore for best research publication, a Scholarly Achievement Award from the Institute of Oriental Philosophy, Tokyo, the Bhoogol Ratna Award and the Gandhi National Fellowship. He has supervised innumerable doctorates and attended countless seminars and conferences; so many that they are not all recorded on his 20-page curriculum vitae. Whether in the capacity of a nominee for selections and examinations or invitation to deliver lectures Misra's reach extends far and wide in the departments of geography in India. Other than geographers, Misra has led research teams of development planners and economists as well as regional scientists. Many of his 50 books are compulsory reading in the curriculums of geography departments across India. India's leading publisher of geography books confirms that his books sell well. How has Misra garnered popularity?

It could be said that Misra's wide range of works gave him a high measure of visibility. Misra's geographical locales of work have stretched far and wide starting with his first appointment as a researcher in the National Atlas and Thematic Mapping Organization, Calcutta; reader in the Department of Geography, University of Mysore; founder of the Institute of Development Studies, University of Mysore; vice director of the United Nations Centre for Regional Development, Japan; vice chancellor, University of Allahabad; professor in the Department of Geography, Delhi School of Economics; chairperson of Gandhi Bhawan at the University of Delhi; and consultant to the United Nations Development Program in a number of countries. But more than the niche he occupied it is the method of anchorage that place him immediately in the spotlight.

Misra survives in a habitat of work not so much by adapting but by changing the environment. Introducing daring alterations, Misra captured immediate renown. Take his appointment at the Department of Geography, Mysore in 1965, in a short ten months he overhauled the postgraduate department and launched a Centre for Urban and Regional Studies. Finding the department too small to contain his ideas, in less than five years Misra set up the Institute of Development Studies. While the city of Mysore acquired one of its meta-disciplinary institutes, for Misra it became a platform for interaction with other social scientists and a centre where novel programmes in name and style were created. One cannot dispute that a master's in development planning, and a diploma in environmental planning as well as multilevel planning, more so through correspondence, were unique to the Indian academic landscape in 1971. Inaugurating the in-house journal *Regional Development Dialogue* in a few weeks after his appointment at the Regional Development Centre. As vice chancellor of the University of Allahabad, he successfully created seven new disciplines.

On a bookshelf Misra's personal publications stand out in bold. He has a ten volume set *Contributions to Indian Geography* (1986), another ten volume series on Nagoya regional development (*Regional Development Series*, 1981), three volumes of *Habitat Asia* (1979), and two volumes on *Million Cities of India* (1998). The row does not end here for there are another two dozen books whose attention is not so much driven by quantity but by diversity in themes. Tackling diverse subjects as vast as agricultural innovation to medical geography, rural area development to local level planning, third world peasantry to conflict resolution, and environmental ethics to research methodology, means a greater catch of audience is added to his basket of popularity.

Though an idea created by Perroux as far back as 1955, Misra's staunch belief in the role of growth poles for drawing India out of its chronic underdevelopment compelled him to write on the theme of regional planning. It is no coincidence that 23 of Misra's books and double the number of his research publications are singularly devoted to the theme of regional planning and development. While these may be sound reasons for his wide acclaim one should not forget that in the field of academics not all among those who gain appointments and write books are honoured and attain popularity. Misra claims that, "a compassionate understanding of events and people, success and failure, needs and wants, is of equal importance if one hopes to carry along a large following." His pragmatism has deep origins.

Misra was born on 5 September 1930 in the village of Hanumanpur in the Pratapgarh district of the state of Uttar Pradesh. At a high school in Kalakankar, the veteran Gandhian, Naresh Bahadur Srivastava evoked in Misra ability in the twin roles of compassionate understanding along with reasoning. Exposed to the writings of Marx and Gandhi, it is under the supervision of his favourite teacher that Misra learnt the first concepts of man and society. Thereafter, as a student at the University of Allahabad and a doctorate at the

University of Maryland Misra's insight about the question of development and its wider ramifications was enhanced. That hybrid learning is essential for comprehending the complex nuances of development, prompted Misra to take a double degree in post-graduation: geography at the University of Allahabad and economics at the University of Agra. Even in the USA, he took courses on agricultural economics and sociology, alongside his doctoral work. More than the books, Misra's pragmatism was honed by his work as an assistant in the Lok Sabha Secretariat of the parliament of India at an impressionable age between 25 to 28 years. Travelling across the country with Members of Parliament, improved his efficiency to produce voluminous reports at short notice. Though brief his encounter with parliament implanted a deep ambition to command and channel people into a field of work. Driven by understanding along with intellectualization, today Misra has acquired popularity among geographers in India. In modesty he adds, "he is not gifted but guided." What is his next forte? At the age of 71 his remaining desire is to enter parliament. His dynamism spurs him to plunge into a study on Gandhi and establish a Gandhi Dham at Allahabad. Honoured with the first National Gandhi Fellowship, Misra is all set to edit a book titled *Rediscovering Gandhi: The Study of Gandhi*. His latest research could take Misra nearer to an indigenous theory of spatial development, it could take him close to a search for truth in life, or even head him straight to the houses of parliament in India. On any or all three fronts, Misra's popularity is bound to increase. (1930–2021)

Aijazuddin Ahmad

Ahmad does not belong to the band of human beings who are easy to comprehend, condense and pen. It is perhaps a truism that simplicity is elusive. In his journey as an academician Ahmad's commitments have been intense, perhaps a little too emotional to be in tune with the academic ethos prevailing today in the portals of universities in India.

Ahmad was born on 12 January 1932 in Firozabad on the agrarian lands of Western Uttar Pradesh. One day his family received a packet wrapped in a map of the United States of America. While his father, an employee for the British Department of Revenue, found the contents urgent, to the ten-year-old Ahmad the wrapping paper was a delight. Developing a love for maps, a book on world deserts, bought with bargained care from a local *kabadi* shop, instilled in him an urge to know the world beyond his little world. Early in life he resolved to learn and research about the world – if geography became a natural choice, the Department of Geography at Aligarh was the logical destination. Carrying the flag of being India's oldest Department of Geography and a close proximity from home were the attractions. Housing one of the finest collections of geographical journals and books, if the Seminar Library was his favourite refuge, it was the fervent intellectual activity in Aligarh which kindled Ahmad's instinct to delve into the causes of wide-scale disparities and continuous human suffering. Vigorous discussions and heated debates became forays for geographical thought; it is in his most private moments that emotions gave birth to poetry and prose. Recording the pains in life, Ahmad's profuse publications in Urdu literature and his geographical contributions cannot be divorced. Gifted with a literary flair, imbibing an understanding of spatial concept, writing was Ahmad's academic companion in years to come.

When formal education came to an end in 1962 with the submission of his doctoral thesis, "Human Geography of the Indian Desert," Ahmed was offered a lectureship at the Department of Geography, Aligarh. Whereas the carefree days of life had kept Ahmad in Aligarh for 18 continuous years, first as a student and later as faculty, a sense of being stifled drove him in 1972 to the more socially relevant geographical research promised at the newly established Centre for the Study of Regional Development of Jawaharlal Nehru University, New Delhi. The core of his interest at the centre revolved around social geography which embraced not just perspectives on historical geography, regional development and planning and demography but also drew heavily from critical geographical thought. His authorship of a dozen books such as *Social Structure and Regional Development* (1993) and *Muslims in India* (1993), along with 60 research papers in national and international journals, capture Ahmad's reflections on different social dimensions. His ideas over a quarter century of teaching and research at the centre found culmination in his recently published book entitled *Social Geography of India* (1999). He now hopes to add a companion volume on critical social geography in India which will address issues of minorities, caste and tribe.

His power of exceptional literary thinking afford high quality to most of Ahmad's publications and has not only won him fellowships to the former Soviet Union Academy of Sciences and the Fulbright Program in the US but has also preoccupied his time as editor of a number of leading journals in India. During his days at Aligarh, he was editor of *The Geographer* (1964–1972), honorary editor of the *Journal of Abstracts and Reviews: Geography (1981–1988)* for the Indian Council of Social Science Research, editor of the *Annals of the National Association of Geographers*, India from 1982 to 1994. Few geographers in India can boast of holding office as editor for so long as Ahmad. Loyal to the tradition of the student–teacher relationship, few would also know that Ahmad was a ghost writer and worked on a number of research projects and papers published either under joint authorship or even without claims to a name.

Ahmad's critical insights, though a boon for intellectual work, became a bane when targeted at mediocrity, which permeates the pool of scholarships in India today. Though he understood the social dynamics of a complex society like India, ironically his ability to deal with the people and processes of his professional environment were beyond his skill. The result is that more critics than friends fill his balance sheet of the day. But then, who said popularity was the dominant measure for scholarship?

It is often said our intentions mirror the world outside. Scholarly passions kept at bay any offers of administrative and management posts and, freed from mundane routines, gave Ahmad ample time to indulge in interests of the finer arts of wide reading, classical music and travelling. Learning from the kaleidoscope of his life experiences, Ahmad, in his post retirement days hopes to one day write his memoirs as a geographer in India. Though the idea is still at an embryonic stage, when complete it would be a blend of realism and idealism punctuated where prose fails with Ahmad's very own lines of poetry. (1932–2005)

Kavasseery Vanchi Sundaram

On 18 August 2000, under the umbrella of geographers, the first consortium of earth scientists was launched. It was christened the Bhoovigyan Vikas Foundation. The occidental and oriental flavour in its name perhaps heralds the spirit of its common search for sustainable lifestyles and sustainable development. Sundaram was not merely the founder chairperson but the moving force behind this professional non-government organization. The mission was to celebrate Earth Day with a pledge to work for a sustainable future, to initiate a lifetime achievement award of Bhoogol Ratna to eminent geographers and set up the first geographical museum in India.

By August 2001, all three objectives had been met. The Earth Day flagged, the Bhoogol Ratna Awards given, and a 300-acre piece of land cordoned for an earth museum. In a few months, Sundaram achieved what many geographers had long talked about – he created a scope for the foundation of an institute of geographical studies. The event reveals two features: first, Sundaram has exceptional determination and second, his devotion to the discipline of geography is in an exceptional niche. The uniqueness comes to the fore when one realizes that in his younger days, Sundaram had been sceptical about pursuing geography as a discipline of study. He spent all his professional life in the company of non-geographers. He never taught at any university department of geography. He never supervised a Master or Doctor of Philosophy on this subject. He was a lecturer in the Alagappa College at the University of Madras from 1951 to 1958. When he resigned from here as a geography faculty, he was 29. He remained active in the domain of planning for no less than four decades. In 1998 he was elected as president of the National Association of Geographers, India. At 69 years he had rejoined the community of geographers. He exemplifies a return of the native.

Sundaram's sojourn of 50 hectic years of professional life involved him in national and international agencies across India and countries of Europe, Asia and Africa. He was one of the first geographers to be appointed to the Town and Country Planning Organization, Delhi in 1958, an office he served for 15 years. Over the next 14 years he rose up the ranks from joint director to joint adviser at the planning commission. During the 1990s Sundaram globe-trotted with assignments under the banner of the United

Nations and other international organizations like the United Nations Centre for Regional Development, UNESCO, the Food and Agricultural Organization, and the Centre on Integrated Rural Development for Asia and the Pacific. In geographical parlance, Sundaram's career has scaled from the micro- to meso-and all the way up to the macro-realms. His mission in life was to deliver the tenets of planning within a spatial context.

The event that propelled Sundaram to discard academics and move to the applied domain of planning was a traumatic one. After an extended village level fieldwork in the Cumbum valley, Sundaram was in the throes of finalizing his doctoral submission when the university authorities informed him that a thesis cannot be accepted from a college with only an affiliating status. The administrative mess frustrated Sundaram to the hilt; dumping his thesis and resigning his teaching post he headed towards Delhi to join the newly created urban planning body. Circulating an extensive questionnaire Sundaram, along with 15 other investigators, surveyed 368 villages around Delhi. Following the premise and postulates of the central place theory, Sundaram drafted the rural development plan for Delhi. When the Town and Country Planning Organization was formally inaugurated in the capital, Sundaram was quickly inducted without hesitation. The First Master Plan of Delhi, and regional plans for Rajasthan Canal Command Area, Dandakaranya, South-East Mineral Industrial Region, among many others carry Sundaram as a member of their team. Gaining valuable experience when the planning commission advertised the post of geographer–planner Sundaram had few equals. It is with this apex planning body of India that Sundaram's concepts and insights as regional planner evolved and matured.

From centrally sponsored schemes to multilevel planning to decentralized planning, Sundaram churned out projects and led study teams. The critical lesson he learnt was that the efficacy of a regional plan needs to be bolstered by the effective training of the officers who are finally responsible to implement the objectives. By the end of the 1980s the climate of thinking was focused on how to remove bottlenecks to achieve equitable spatial development. Sundaram specialized in the role of translating goals into reality in the area of rural development. The innovative training modules which he designed attracted attention from international development agencies. Sundaram's wisdom, seasoned by issues of regional planning and his insightful understanding of ground realities, made him a much sought after consultant with the United Nations. Over a dozen prestigious assignments stud his professional life from 1987 to 1995. Forwarding a voluntary retirement to the planning commission, Sundaram spread his advice on issues of regional planning. Many areas of development in countries like Nepal, Bangladesh, Thailand and countries in sub-Saharan Africa hinge on the manuals and plans spelt by Sundaram. Seventeen reports, many running into several volumes, and over 120 technical papers, decorate the bookshelves at his residence in Delhi. After a profession that was remote and distant from the academic mainstream, how was Sundaram able to return as a native?

Born on 30 November 1929 in Calicut to an orthodox Brahmin family which was overtly conscious about education meant that Sundaram inherited the urge for a literary life. It is no surprise that after a cursory attempt to acquire a doctorate, Sundaram nursed back his wounds only after he acquired a Doctor of Literature degree. To confirm the pathbreaking research of his thesis, the university sought reviews from over a dozen different referees. It is this collective praise that metamorphosed into a Doctor of Literature, the first of its kind given to any social scientist by the University of Mysore in 1977. The experience in planning forever kindled in Sundaram the desire to pen his learning. His book *Urban and Regional Planning in India* was published in 1977. This book is his jewel in the crown and won him the Nuffield scholarship for post-doctoral work at the University of Cambridge in 1978–1979. By the end of the 1990s, Sundaram had written 11 books and contributed 27 chapters to different books. The tenor of his work is evident from the fact that nearly half the books and 90 per cent of his articles are authored by him alone. That he does not dabble superficially into themes is proven by the fact that a major share of his publications square on issues of regional planning and development alone. Sundaram's publications find multiple entry in the curriculum of leading departments of geography, and serve as valuable references on courses in development and planning.

All along as Sundaram soared the skies to become a national and international consultant on regional planning, he did not forget to nourish the roots of his discipline of geography with books and publications. Launching an elaborate blueprint for the Foundation for Propagation of Earth Science is an asset that makes Sundaram's contribution to geography special. His return was more than welcome! (1929– 2013)

Gur Bachan Singh

An allure for travel, a penchant for photography and hospitality, the bold steps of approaching authorities, and the desire to experiment and lead all typify Singh as a person. His long innings of teaching for 35 years was nourished by his love for the land and people in a rural setting. Tilling his research with first-hand experience drawn from his background in a farming family, his work on agricultural geography reached its zenith through ten years of labor and over ten hours of daily toil in the *Agricultural Planning Atlas of Punjab*. While the inspiration for this atlas was instilled by Copock, his doctoral supervisor at the University of Edinburgh, what adds novelty to Singh's atlas is the dense descriptive material in narrative and tabular form, in both English and Punjabi. His innovative bilingual methodology made the atlas a model for different states of India. Winning Singh praise and recognition, it proves that even a single solid piece of research can be most rewarding.

Territorially, Singh is deeply rooted in the soil of Punjab. His entire education was there and he was chairperson of the Department of Geography at Punjabi University, Patiala. All 12 of his books and research papers are on the state of Punjab, 18 of the 38 conferences and seminars he attended and 14 of the 21 honours or posts conferred on him are from the state. He supervised three doctorates and 20 master of philosophy students, all of whom researched on Punjab. Except for the years he spent at Edinburgh, UK with the Lamb Fund Award 1973, his entire education was drawn from Punjab. Singh was born on 2 August 1936, to a farming family in the village of Ganaur in the district of Gujranwala, now in Pakistan Punjab. The family moved to Indian Punjab after the partition of the country in 1947. More than a mere affection for his birthplace, Singh's addiction for Punjab stems from the fact it is one of the most prosperous and rapidly progressing states of India. The opportunities it provided to Singh as a geographer were far and wide.

A practical man, early in his professional career he tried to solve problems. Identifying a pressing need of geography texts in Punjabi, he translated books on diversified fields in geography such as, world geography, physical geography and economic geography of India at an early age of 27 years. His textbook, *Social Studies in Punjabi*, for the fourth and fifth standards remains a model text for nearly four lakh school students under the Punjab Education Board. As member of the Punjab Flood Control Board and the Border Area Technical Committee and president of the Punjab Heritage Society his advice on various issues found applied value. At his suggestion tunnels were constructed at certain key locations along the rivers to prevent the havoc caused by flooding in these areas; its success proved that geographical solutions are sustainable. Compiling a four-volume *Encyclopaedia of Punjab Heritage* is Singh's latest passion. Capturing through the lens the deteriorated condition of monuments in seven districts scattered over the Majha, Doaba and Malwa regions of Punjab, Singh hopes to draw the attention of policymakers towards the restoration and conservation of these sites.

Today Singh is giving his atlas a new shape. Just two days after retiring from the Department of Geography, Punjabi University, Patiala, Singh delivered his presidential address, "Green Revolution: Gains and Pains" at the 21st Indian Geography Congress, held at Nagpur in 2000.

Despite a track of nearly half a dozen presidencies in different societies and associations, such as the Association of Punjab Geographers, the Punjab Heritage Society and Kalgi Dhar Gurdwara, he claims that of the National Association of Geographers, India was indeed a prized honour. He insists that the presidentship of NAGI was a call from the divine. His curriculum vitae records his presence in the 3rd, 8th, 16th, 20th, 21st and 22nd Indian Geography Congress. Each congress was a landmark in his life. At the third congress held at the Centre for the Study of Regional Development, Jawaharlal Nehru University, New Delhi, he had a first-hand encounter to meet some of the most famed Indian geographers. His atlas received an academic release at the 8th congress held at Srinagar. The 16th congress held at Hyderabad conferred on him a citation for his distinguished and outstanding contribution in the field of agricultural geography during his illustrious career devoted to the cause of the promotion and development of geography. He was elected president at the 20th congress held at Gorakhpur. Singh's life as a geographer blends a happy light-hearted mix of work and pleasure. (1936–)

Gopal Krishan

Gopal Krishan's urge to learn is insatiable. Be the experiments to experience being a leader or follower, the pursuit of a single or multiple theme, specializing in one or generalizing in numerous areas, for him the fire to expand the vistas of knowledge had neither a route nor a routine. The familiar tag GK – an acronym which friends and students use interchangeably to address him as a person or one with 'General Knowledge' probably bears compliment to his personality. From teaching to research, to consultancy, from projects to plans to people's programmes, articles and books, book reviews to atlas and a blueprint for a geographical museum are part and parcel of his work. With assignments at regional to international universities, as president of India's largest geographical and cartographic associations, interacting with economists, sociologists, architects, environmentalists and planners, GK's work is a dense mosaic of uncommon diversity. From the Green Revolution to rural transformation, regional disparities to regional identity, rural–urban relations to pricing of urban water supply, hill towns to our national borders – GK's search spans a great variety.

His academic canvas is akin to a modern art painting. Transgressing bounds or boundaries the umpteen colours, shapes and form defy a ready-made or simple caption. Careful observation decodes a heartfelt message: to make geography a more relevant social science.

GK's impulse to learn and search for relevance is manifest in the evolution of his research interests. Inheriting a legacy of population studies from his mentor, Professor Gosal, at the Department of Geography, Panjab University, Chandigarh, he was quick to realize that research confined to the census of teeming millions would be meaningful only if integrated with an understanding of the development process. This explains his alert decision to take on development studies in the 1980s. While his keen interest was to tackle issues of development at both regional and national levels, he elected urban areas as the microcosm for analysis and application. Population, development and urban studies are the dominant strokes of his brush. Whether by design or accident all three themes have a near equal one-third slice of his 103 research papers, 30 on development, 33 on urban and 32 on population with the remainder falling into a miscellaneous category. What is remarkable about the list of writings is not simply the numbers, nor the large topics that catalyse these three themes, but the fact that no area of his research was abandoned as the range of his interest expanded. His canvas, though diverse, is deftly welded in unison.

An enthusiastic learner, he galloped to first-class grades and won himself ranks and scholarships all through school, graduation and post-graduation. His outstanding scholastic acumen attracted attention, and at the age of 22 he was drafted into the Department of Geography, Panjab University, Chandigarh. Nestled at the base of the Jakhu Hill, it is in the non-elite and down to earth DAV School of the Himalayan town of Shimla that GK's desire to learn was kindled. To enrol in the Indian Administrative Services had been his first dream but this he did not pursue. Teaching was a second career option for GK. Reticent and reclusive by temperament, books were a retreat which offered much and asked little in return. The play of luck and the joy of being appointed as a lecturer so young, at one's own alma mater, anchored GK to Chandigarh for a lifetime. Despite his 38 years of teaching, the pent-up desire to serve the needs of the country and interact with administration compelled him to step out of his academic camp.

Invitations to teach and research gave GK prestigious appointments not only in India as a professor of Regional Planning, at the National Institute of Urban Affairs, Delhi but also in leading universities overseas. The Young Social Scientist Exchange Programme took him to Liverpool University, the Indo-British seminar to a number of leading centres, while it was an award from the Royal Geographical Society, London that gave him the opportunity to join the University of Cambridge. As a guest professor he visited Bemidji State University in the USA, the University of Bremen in Germany, and the University of Pecs in Hungary.

From these experiences GK brought home many nostalgic moments and fond friendships. Bonds were fostered with Prothero, Toth, Farmer, Dickason and Bahrenberg, among others. A rare honour for GK was the degree of Doctor of Science conferred by the University of Pecs in 1996 on the occasion of the millennium celebrations of Hungary, of its birth as a nation in 996 AD. Perhaps more than these tangible benefits are the fertile ideas which GK carried back. Interactions with Chisholm at Cambridge for example had strengthened his interest in administrative geography; GK wanted to push administration as a separate branch of study in geography in India. His failure to meet success in this direction was perhaps less to do with the traditional mind set of geographers in India and more with the fact that GK himself was unable to produce a valuable and simple text for easy reference on the theme! The desire to embrace realms larger than that of the university thwarted the publication of such a text but on the flip side enriched the centres of application where he took an assignment.

A member and consultant to the Punjab State Planning Board, the National Institute of Urban Affairs, the Human Settlement Management Institute, and the Central Tibetan Administration, and taking projects sponsored by diverse organizations such as the Indian Council of Social Science Research, the Institute of Rural Management, the University Grants Commission, and Chandigarh Administration allowed GK to research and take findings to people who he thought would act upon them. His book, *Spatial Dimensions of Unemployment and Under-Employment* gained wide currency. Another among his diverse ventures: *Inner Spaces – Outer Spaces of a Planned City: A Thematic Atlas of Chandigarh*, serves as a crucial compendium for research, planning and administrative decision.

It is these diversified experiences and assignments with a range of universities, disciplines and institutes that fill GK's bag to the brim with 103 research papers, 19 projects, nine books, three monographs, half a dozen book reviews, and the supervision of ten doctorates and 22 masters of philosophy. Working in meta-disciplinary groups is what strengthens GK's belief that research is spruced up when worked in a team. It is little wonder then that half his research projects and a quarter of his publications are co-authored works. That his research papers span 14 different journals, both national and international, speaks volumes of the quality and range and breadth of his themes.

Within this dizzy pace of consultancy and publications, GK's true love remains the Department of Geography at Chandigarh. The tenor and diversity of his contributions spill into administrative duties. Artfully advocating the merits of the department to the University Grants Commission, GK and his colleagues were able to attract the much sought after prestigious Special Assistance Program from the University Grants Commission not once but twice. Under the umbrella of this programme GK, as coordinator, was able to not only refurbish the department with state-of-the-art machines but also enthused a vortex of activities like seminars, workshops and invitations to visiting faculty to enrich the corporate life of students and faculty.

Ever filled with plans, he is all set to roll the ball as Chairman of the Editorial Committee for Fifth Survey of Research in Geography under the auspices of the Indian Council of Social Science Research in Geography. Perhaps closer to heart is his dream to consolidate experiments and experiences, projects and monographs, teaching and research into a book titled, *Sustainable India: A Geographical Perspective*. To conquer domains and soar to new heights of excellence; Gopal Krishan's urge to learn does sustain. (1940–)

Bibliography

Ahmed, Aijazuddin. 1999. *Social Geography of India*. New Delhi: Rawat Publications.

Alam, Shah Manzoor. 1965. *Hyderabad – Secunderabad – A Study in Urban Geography*. Bombay: Allied Publishers.

Alam, Shah Manzoor (ed.). 1976. *Planning Atlas of Andhra Pradesh*. Pilot Map Production Plant, Survey of India.

Dayal, Parmehswar. 1974. *Encyclopaedia Britannica* (15th ed.). https://www.britannica.com/contributor/P-Dayal/709

Harris, Chauncy Dennison and Edward Louis Ullman. 1945. "The Nature of Cities." *Annals of the American Academy of Political and Social Science* 242: 7–17.

Hoyt, H. 1939. *The Structure and Growth of Residential Neighborhoods in American Cities*. Washington, DC: Government Printing Office.

Krishan, Gopal. 1986. *Spatial Dimensions of Unemployment and Under-Employment: A Case Study of Rural Punjab*. New Delhi: Concept Publications.

Krishan, Gopal. 1999. *Inner Spaces – Outer Spaces of a Planned City: Thematic Atlas of Chandigarh, Chandigarh Perspectives.* Chandigarh: Peco Printing Press, Industrial Area.

Misra, Rameshwar Prasad. 1979. *Habitat Asia: Issues and Responses* (Vols. 1, 2 & 3). New Delhi: Concept Publications.

Misra, Rameshwar Prasad. 1981. *Regional Development Series.* Maruzen Asia on behalf of the UNCRD Japan.

Misra, Rameshwar Prasad. 1986. *Contributions to Indian Geography.* South Asia Books.

Misra, Rameshwar Prasad. 1998. *Million Cities of India: Growth Dynamics, Internal Structure, Quality of Life and Planning Perspectives.* New Delhi: Sustainable Development Foundation.

Ram, Laxmi Niwas. 1991. *A Systematic Geography of Bihar.* Geographical Society, Department of Geography, Patna University.

Ramakrishnan, Moni and Rajagopala Vaidyanadhan (eds.) 2008. *Geology of India* (Vol. 1, p. 556). Geological Society of India.

Ramakrishnan, Moni and Ajagopala Vaidyanadhan (eds.) 2010. *Geology of India* (Vol. 2, p. 428).

Index